RESTLESS SOULS

itbooks
AN IMPRINT OF HARPERCOLLINS PUBLISHERS

Restless Souls

The Sharon Tate Family's Account of Stardom,
the Manson Murders, and a Crusade for Justice

Alisa Statman

with **Brie Tate**

FIRST IT BOOKS PAPERBACK PUBLISHED 2013.

Designed by Paula Russell Szafranski

The Library of Congress has catalogued the hardcover edition as follows:

Statman, Alisa.
  Restless souls : the Sharon Tate family's account of stardom, the Manson murders, and a crusade for justice / Alisa Statman, with Brie Tate. — 1st ed.
    p. cm.
  ISBN 978-0-06-200804-6
  1. Tate, Sharon, 1943–1969. 2. Manson, Charles, 1934– 3. Murder victims' families—United States—Case studies. 4. Actresses—California—Los Angeles—Death—Case studies. 5. Murder—California—Los Angeles—Case studies. 6. Criminal justice, Administration of—United States—Case studies. I. Tate, Brie. II. Title.
  HV6534.L7S73 2012
  364.152'3092—dc23   2011040364

ISBN 978-0-06-200805-3 (pbk.)

19 20 21 22 DIX/LSC 10 9 8 7 6 5

This book is not intended to harm anyone in the Tate family's past. In sharing these experiences with readers, the intent is to foster awareness of the impact crime has on victims' families and the rights the Tate family fought hard to establish and maintain. In a few places, we have changed or left out entirely the names of individuals who have played a key role in these events in order to preserve their privacy. We have also, in some places, altered details, locales, and other specifics to be sure these people are not recognizable, but in no instance have we altered or changed the stories entrusted to us and shared here with you.

To Patti,

P.J.,

Gwen,

Sharon, and her baby.

With much love,

A.S. and B.T.

Contents

Preface

Sharon Tate is a name that most have bound to the word *murder*. But, for me, my aunt's name inspires thoughts of strength, determination, courage, and unconditional love. Sharon touched many during her lifetime with those traits. With her death, she then touched so many more in the limitless way that tragedies do. My grandparents, Paul and Doris Gwendolyn Tate, ingrained those attributes in Sharon during her formative years, and in the aftermath of her murder they fortified those same traits in my mother, Patti Tate, and then in me.

My grandmother, in her infinite wisdom, realized early on that we learn and grow most from the heartaches life tends to bring us. Through the infamy of Sharon's murder, she was given a platform that could reach millions, a voice that could bring about positive change in the world. With this gift, she dedicated the rest of her life to helping others.

After my grandmother died, my mother went through a major transformation. Her life motto until that point was: DON'T ROCK THE BOAT. But my grandmother's death sparked in her a need to stop being scared and to start living her life. After that, she picked up the torch my grandmother had carried before her, advocating for victims' rights and fighting to keep Sharon's killers imprisoned.

I was very young when my grandparents and mother died. By the time I was mature enough to care about my family's history, it seemed there was no one left to help with my mounting need to connect with my roots. But then I came across my mother's unpublished autobiography, which she and Alisa Statman wrote years ago. Within the pages many of my questions were answered and ultimately it revealed to me why my mother and grandparents had spent much of their lives as restless souls. Years later, with Alisa's intimate knowledge of my family, my mother's autobiography was reworked into this family memoir.

Because of what my mother and grandparents instilled in me, I live my life daily fighting for what is right, not always with the ambition and determination that they did, but with love in my heart to carry forward their positive force. By publishing this book and sharing our story, I hope that readers will prosper from our experience, and perhaps in my own little way I will also help others by bringing about a bit of peace to the restless souls who follow our journey.

—Brie Tate

Introduction

It seems natural to ask Why this book? Why now? The truth is, this book has been brewing within the Tate family since 1971, when Sharon Tate's father, P.J., attempted to write his autobiography. In addition to his efforts, Sharon's mother, Doris, also had numerous irons in the publishing fire. Sadly, though, neither memoir was ever published, as each passed away before their books were completed.

A generation later, Sharon's sister Patti and I moved into her childhood home as domestic partners along with two of her three children—Brie, who was nine years old, and Bryce, who was just six. Over the course of the next few years, Patti's frustration over the constant inaccuracies and misportrayal of her family in the media peaked, so together we wrote her autobiography in an attempt to ease her mind, set the record straight, and fulfill her parents' goal of sharing their important personal stories as well. But after breast cancer took Patti from us, plans to publish were laid aside as we all grieved our loss.

In the years following Patti's death, Brie and Bryce stayed in my custody. With the support of Patti's father, P.J., we remained in the Tate family home. In time, the kids grew into young adults. I had a few more wrinkles around the eyes. And Patti's autobiography had gathered a lot of dust until the day Brie read it. Shortly after turning the last page of the manuscript, she plopped it down on my desk and said, "You have to try to get this out there so people can read it."

I explained to her that with her mother dead, there was little hope that it would see the light of day. With a willfulness definitely passed down from her mother and grandmother, Brie pushed the pages closer to my hands and said, "I'm sure you'll figure something out."

That something eluded me until P.J. decided to sell the family home to me and move to Whidbey Island for some peace and quiet. He packed his few remaining personal items from the house and left the rest with me to do with it as I saw fit.

While P.J. boarded a plane for that faraway island, I was left to the task of cleaning out and organizing our home—a home that had accumulated the lifespans of three generations. Room by room, box by box, I unearthed a treasure of the Tate family's home movies, audio and video recordings, journals, and letters dating back to 1961, as well as a massive archive of police and court documents from Sharon's murder case.

It took months to catalog it all, and during that period I was pulled into a time warp of their formidable lives. As I closed in on finishing the task, the answer that had earlier eluded me now seemed clear. I could expand Patti's autobiography into a family memoir. By using what I'd found, as well as the personal knowledge shared over the years with me by P.J., Doris, Patti, and Brie, I decided to write it from each of their unique and extraordinary perspectives.

By combing through all that information and then reconstructing the work into a four-decade, cohesive narrative, my goal was to chronicle their lives with historical accuracy in even the finest details. Nevertheless, with four of the five key witnesses to this story gone, there were a few times when I was left to fill in the gaps with my personal interpretation.

Police, prosecutors, and defense attorneys alike will concur: There is no perfect witness. When the last page of this book has been turned, some will agree that I am certainly no exception to that rule. Over the years, friends and relatives have shared valuable anecdotes with me. But when those stories varied from Patti's, P.J.'s, Doris's, and Brie's, I left them behind in order to preserve and honor the shared memories of the loving individuals who are the heart and soul of this book. . . . Now, after three generations' time, this is finally their story.

—*Alisa Statman*

RESTLESS SOULS

THE FIRST DAY OF THE REST OF OUR LIVES

I left my sister's house one night . . . and life was good.
Then I woke up to another day and life had changed so
very, very dramatically as our world just fell apart and I
realized that it's never ever going to be the same.

—PATTI TATE

Patti   August 9, 1969

"My God, Sharon's been murdered." Barely able to get the words out, my mother collapsed against the scarred door frame and then to her knees. I looked up from my favorite cartoon in time to see the first tear spill from her eyes.

Paralyzed by her emotion but not understanding it, I could only stare at her while the seconds passed, waiting for an explanation. Her lips fluttered, but there was no sound. Leaning forward, I strained to hear. Then, in a scarcely audible whisper, she said, "My baby's dead."

As if floating to me in delayed time and space, her words eventually reached my ears, forever altering the stability of my life.

The morning hours preceding that moment began as so many others had in my eleven years, with only Mom and me in the kitchen. My oldest sister, Sharon, moved out years ago, Dad, an army intelligence officer, was stationed in San Francisco, and my sister Debbie was hibernating in her room because despite how mature she felt, we were the enemy who reminded her she was only sixteen. Unlike my relationship with Sharon, I felt distanced from Debbie. She was too young to be a role model, and too old to be a friend.

With the clang of the last dish placed on the breakfast table, the phone rang. Assuming it was Sharon, I nudged my ear next to Mom's. "Doris, have you talked to Sharon this morning?" the voice asked.

I felt Mom's body stiffen. "Why?" she asked.

"Turn on the radio. They're reporting some trouble in Benedict Canyon."

"What kind of trouble?"

Thoughtful silence, then, "I'm not sure. I have to run." The click of disconnection.

Just north of Beverly Hills, Benedict Canyon winds its way through the Santa Monica Mountains in a labyrinth of entangled and unmarked roads. A quarter of a mile before Benedict Canyon Road ascends the mountain, is Cielo Drive, a narrow road that abruptly climbs the hillside. Without the benefit of a street sign, Sharon's cul-de-sac is an inconspicuous left turn off the main path. Though her address is 10050 Cielo, it is often confused as an outstretch of Bella Drive, the marked road directly opposite.

Always one to overreact with worry, Mom simultaneously turned on the radio and dialed Sharon's number.

On the radio, a newscaster was mid-story: "Reports of a possible fire or landslide first came over the police-band radios at 8:30 this morning. Our correspondent at the site reports that at least three people have perished in this disaster on Bella Drive in the Benedict Canyon area. Police are supplying little information at this point, and are withholding the victims' names pending family notification. We'll be reporting on this throughout the morning as the information comes in."

Bella Drive sounded familiar. "Isn't that where the lady with the good cookies lives?" I asked.

I noticed a slight tremble in Mom's hand as she lit her second Tareyton in ten minutes. "Uh-huh," she absently responded.

It turned out that the lady with the good cookies was Doris Duke, heiress to the American Tobacco Company. But when I met her, I measured a person's importance on the type of cookies they served.

February 14, 1969, was Valentine's Day, and the day that Sharon moved into the Cielo house. Two weeks later, Sharon was still unpacking boxes when I noticed the backside of the Duke estate across the ravine. Intrigued by its resemblance to a castle, I hiked to the front of the property. An engraved plaque to the right of the open gates read FALCON'S LAIR. Sharon had mentioned Falcon's Lair when she talked about a haunted house in her neighborhood! Fearlessly, I started through the gates and down the sloped driveway in search of a ghost named Rudy.

Ten feet into my adventure, there was a voice, English and bellowing. "This is private property. What are you doing here?"

I wheeled, losing my footing on the gravel drive. Just then, a long, black limousine arrived, distracting me from the red beading across my knee. The car glided to a stop when the back door was even with us. The darkened window slid down. The woman behind the glass curiously looked at me, her eyes above lowered sunglasses. "Who do we have here?" she asked the man.

"My name is Patti," I blurted before he had a chance to answer. "My sister lives across the way in the big red barn."

The woman's eyes shifted between us as if we were crime partners, eventually settling on the man. She pushed her sunglasses back in place. "Stop being such an ogre and bring Patti in so we can clean those scrapes. And get me the Polanskis' phone number."

Before I'd even had a chance to take in the royal surroundings, my knee was bandaged, tea and cookies tantalized from the coffee table, and Sharon's arrival was announced. "Mrs. Polanski here to see you, Ma'am."

Polanski was such an odd name, and even though Sharon had married director Roman Polanski over a year ago, she was still a Tate in my mind.

Until I saw her nervously chewing her lower lip, it hadn't occurred to me that I'd done anything wrong. Her hand reached out. "Miss Duke, I'm terribly sorry for the intrusion."

"Nonsense," she said. "We're having tea. Please join us."

The sun streamed through the kitchen window and reflected off Mom's glasses. I couldn't see her eyes, which always revealed her mood. "I hope Miss Duke is all right," I said, adjusting my view and looking for one of her reassuring winks.

With the phone cradled on her shoulder, she was too preoccupied to respond. No one answered at Sharon's house. I trailed Mom's gaze to the clock and knew what she was thinking. Almost eight months into her pregnancy, Sharon slept late into the morning. It was too early for her to be out of bed let alone out of the house.

With her biscuits and gravy still untouched, Mom made a series of phone calls beginning with Sharon's obstetrician, then onto friends, the police, hospitals, and finally Roman, who was in London. And still, Sharon's whereabouts remained a mystery.

Mom's pudgy fingers pulled at her dark curls until they were limply fringed around her face, all the while her lips murmuring her limited options. If her car wasn't in the repair shop, she'd be halfway to Cielo by now.

I giggled over the reason we were without a car. In three years, Sharon had totaled two cars; everyone—except her—knew she was a terrible driver. A few days ago, Sharon and I had gotten into her car for a trip to the market. *The Ferrari roared to life when Sharon turned the key. Mom waved good-bye from the porch. I waved back from the passenger seat. Sharon throttled the accelerator, shifted into reverse, and off we went—right into the side of Mom's Corvair.*

"It's not funny," Mom scolded halfheartedly. "And don't you let it slip to Daddy—or Roman for that matter. Both of them will pitch a fit that'll last till doomsday."

We finished breakfast with only the sound of the overhead fan's *whoosh, whoosh, whoosh.* With each pass of the blades, Mom's unease rose until it threatened to catch in the whirlwind above. She had a habit of wringing a Kleenex when she was nervous. By 10:30, the one in front of her was pulp. Sharon's routine call was a half hour late.

Before marrying Roman, Sharon had been engaged to Jay Sebring. After their breakup, Jay remained a family friend. His house was a five-minute drive, to Sharon's. "Why don't you call Jay?" I asked.

"I already did. He wasn't home," she mumbled.

I sighed. Ever since Sharon moved out of the house and began her acting career, Mom seemed to be expecting tragedy to strike Sharon. It was just last week that she'd decided Sharon looked too pale. The doctor assured her everything was fine, but Mom insisted on the proof of a blood test.

Proof. Today I didn't have any, so I tried Sharon's tactic—distraction. In the afternoon, we were going to Sharon's baby shower. I gathered wrapping supplies and piled them on the table along with my gift. "Have you wrapped your present yet?" I asked.

"No." She turned from the soap-bubbled sink. "You're giving up Huggles?"

"Yep," I said, more bravely than I felt. "He needs a new baby to protect." I wrapped the worn bear as carefully as china. Sharon had given me my lifelong companion and I was going to miss him, but he was the only gift I could afford.

I placed the bow on top of Huggles's well-wrapped head and propped him up on the counter where Mom leaned in meditation, possibly prayer. Another brew of coffee percolated with a thunderous rumble. So much for distraction. Back to proof. "They said the fire was on Bella Drive not Cielo," I reminded her. "Why are you so worried?"

"Because I think the reporters are confused. There are only two houses on Bella. They would have mentioned Falcon's Lair by now because Doris Duke doesn't have any next-of-kin to notify." Her sentence was choked off by a sob; a sob sure to become contagious if I hung around her any longer.

I grabbed some Pop Tarts from the pantry. "I'm going to watch cartoons."

Without the buffer of a door, it was hard to ignore the eleven o'clock newscast blaring from the kitchen radio. The reporter's voice was more animated than the cartoon I muted. "Two hours ago, police cars raced through Benedict Canyon, responding to what was originally thought to be a landslide. We have just learned that the incident is being assigned to Robbery/Homicide. It is believed that there may be as many as five casualties. One name we've heard repeated over the police radio is celebrity hairstylist Jay Sebring. Whether or not he's a victim has yet to be disclosed. In other news this morning—"

After Mom shut off the radio, the house felt creepy. Why were the police talking about Jay? Had he done something wrong? I expected a reaction from Mom, but the silence held fast until the grandfather clock struck a new hour. The bong of the chime echoed all around. I leaned forward, peering around the corner. She sat motionless; her eyes were wide and fixed on the radio. "Mom?" I softly called out.

Mom was a worrier, no doubt, but this time there was something in her attitude that I'd never seen before. Maybe it was the quiet, or perhaps the fright in her eyes, but it scared me.

Lately, I had been making every effort to be included in the adult realm. In light of the unfolding events, I was content to revert back to the role of a naïve child. I nestled deeper into the cushions.

Each Saturday morning there was a prelude to the shows. "Looking for a magical morning place to be? Step inside the magical world of ABC." *Fantastic Voyage* came on, luring me to the doorway of escape, which I gladly stepped through.

The show was only half over before Mom's panicking voice tugged me from Cartoonville and plopped me back into reality. She was arguing on the telephone with Sandy Tennant; her husband, Bill, was Sharon and Roman's manager. "It will only take Bill five minutes to drive over there. Please. Jay's name came up in the news report, and no one has heard from Sharon since yesterday." There was a long pause. Her tone changed. "He plays tennis every day, dammit! I'm only asking for fifteen minutes of his time." A hushed spell. Then her voice slightly softened. "Will you have him call me as soon as he gets there?"

At noon, Bill Tennant called with the unbelievable news. Mom grasped the phone to her chest as if it were her last link to Sharon.

"My baby's dead." The confusing words rattled in my mind. Murder was not part of my vocabulary. In my experience, death came only to people who were sick or old. How could Sharon be dead?

While Debbie went to get a neighbor, I hugged Mom. "It's okay. Please stop crying. Let's go sit on the couch."

She refused to budge. She refused to let go of the phone. With a fisted hand, she pounded her leg as if to punish herself. "It can't be true. . . . Oh, God, let her be okay," she cried.

My stomach hitched as the first teardrop ran down my cheek. In a moment's time, my life crashed to a halt, yet the familiar music from *George of the Jungle* sang out from the television. If Sharon really was dead, how could the cartoon play as if everything was fine?

Trailing Debbie, our neighbor, Joan, rushed through the door. "Girls, help me get your mama to the bedroom," she said.

The familiar comfort of my parents' room now seemed dark with the fear of the unknown. The sunshine of the day blocked by the shades that Mom never had a chance to open. Joan sat next to Mom on the bed. "Doris, what happened to Sharon?"

Mom's hands masked her eyes. "I don't know. . . . God help me, this can't be happening." Then she calmed, listening to something we couldn't hear. Seconds passed. She bolted from the pillow. "I need to go to her, she needs me."

"No, no, just lay down until we can figure this all out," Joan said, as she gently pushed her back.

Although Mom's head rested, she had not been subdued. "Where's your father?" Joan asked me.

While Joan tried to reach Dad by phone, Mom curled up with her arms locked around a pillow. My mom, my pillar of strength, was crumbling, slipping away from me. I wrapped my arms around her and held on while my eyes snapped shut, willing away the present, praying for the past, and wondering if I would ever see Sharon again. From the other room I heard Joan; "This *is* an emergency. I think there's been a death in the family."

7

My father, Lt. Col. Paul James (P.J.) Tate of the United States Army Intelligence Division, was stationed at the Fort Baker Military Base in San Francisco. The Los Angeles Police had reached him an hour before Joan made the call.

P.J. August 9, 1969 2:00 P.M.

I didn't believe a nickel on the ground until it was in my palm. Getting to Sharon's house was my priority. A coroner's wagon passed as I made a left onto Cielo. My hands ached from my iron grip on the steering wheel. I stretched my equally aching jaw. This had to be a case of mistaken identity. Reasonably, Sharon could have spent the night at a friend's, as was her bias when Roman was out of town. I had every right to expect to find her sitting in the living room, teary-eyed and scared, but alive.

At least two dozen police cars and as many news vehicles clogged the road to Sharon's gate, forcing me to walk up the once familiar cul-de-sac, now alien with their traffic. Along came the whoop of a siren. A cop tried to clear a path. "Go on home, folks. There's nothing to see here."

"Save your breath," I muttered. This crowd isn't going anywhere until their curiosity is satisfied. I passed face after face, the old, the young, a couple holding hands, a child on his father's shoulders, eyes shaded, straining on tiptoes, expectantly waiting. Why are we so eager to view tragedy? Pulling a cigar from my pocket, I pondered the question with the strike of a match. Drawing on the sweet tobacco, I decided it makes us feel more alive.

I stopped near Sharon's gate, bracing myself, yet angling toward delusive faith.

The men at Fort Baker thought I was oblivious to the chitchat around the base. They called me Ice Cube. I prefer being called venerable, though theirs is not a bad analogy. The ice cube is an ever-evolving substance. As individual as a snowflake. For better or worse, everything it contacts is affected—and vice versa. Exposed to shock, one can shatter—and that's just what happened with the LAPD's earlier phone call. "Could you tell me where you were last night?" the faceless Lt. Bob Helder asked me.

"No. My activities are classified."

"Well, can someone vouch to your whereabouts?" Helder pressed.

"I'll bring a note from the principal. You can expect my arrival at thirteen hundred." I slammed the phone to its cradle.

Life plays funny tricks; I was so livid at the detective's accusations that I rode that wave to avoid the undertow of Sharon's fate.

A gate sealed the news teams from Sharon's property. With cameras and microphones jammed against the fence, they shouted questions at the police on the opposite side. My call for an officer's attention melted in with the other voices. Screw it; I pressed the gate control button that caused it to swing away from the frenzy. I pushed with the best of them to the front, until an officer intervened. I held my identification inches from the cop's face until he stood down. A reporter pawed at my shoulder, then shoved a microphone into kissing range. "What's your connection to the murders, sir?"

I swatted the mic away. "If you don't let go of me, son, I'll give you a connection you won't soon forget."

Out of the media's earshot, I turned the tables on the pubescent officer, firing out questions. Army policy: Throw them off guard, get the upper hand.

"Whoa there, sir, you really need to talk with the detective in charge. Just hold tight for a minute."

The second the rookie turned his back, I proceeded up the driveway. The scene stimulated wartime memories. Men combed the hillsides, bushes, even the rooftop of the house. Helicopters intermittently circled overhead. From the opposite side of the canyon, the curious spied like enemy troops.

Mixed between the police cruisers, I saw Jay Sebring's Porsche and three cars I didn't recognize, a Firebird, a Camaro, and a Rambler. The sedan was closer; I peered through the open door. A bloody sheet was slung over the driver's side. I inventoried Sharon's friends, wondering whose car it was, and more important, whose body had been under that sheet?

"Mr. Tate?"

I turned toward the voice.

"We spoke earlier. I'm Lt. Bob Helder. Like I said on the phone, we've already gotten the positive identification on your daughter's body. There's really no need for you to be here."

"Nothing's positive here until I've seen her."

Helder buried his hands in his pockets. "The coroner has removed all but one of the bodies, and the one remaining isn't your daughter's."

I looked up. *"Bodies?"*

Helder nodded. "Five. Bill Tennant made the identifications." He reached into his jacket pocket to pull out a pad of paper and reading glasses. "He identified your daughter, Jay Sebring, Abigail (Gibbie) Folger, and Woo, Woy—"

"That's pronounced *Voy*tek, spelled with a *W*."

"Right, Woytek Frykowski. Uh, we have one unidentified male we found in that car. Any idea who owns it?"

Struggling for composure, I dared not speak. I just shook my head. Three wars had made violent death an intimate enemy for which I'd never shed a tear. Those lives were lost for a reason, and when I made a death notification, I reassured a father that his son died for a cause, always assuming the speech provided comfort. What could Helder tell me? My fingertip blotted the unfamiliar moisture from my eyes. I moved away from him to hide my weakness.

"Mr. Tate, hold on," Helder called.

I couldn't remember how long it had been since anyone addressed me as *mister*, instead of *Colonel* or *sir*. A sinking feeling of drastic change squeezed my heart, clumped in my stomach. Clearly out of my element, I went to familiar territory, an image of a briefing officer apprising me of a mission. *Colonel, your role in this operation will be that of the father. Of necessity, you'll handle this differently than usual. Play it with dignity. No threats, no speeches. The only thing you may find particularly onerous is going home and explaining all this to your wife and daughters.*

Helder said, "We have everything in hand. The best thing you can do for everyone concerned is to go home to your family."

I looked around. "About the only thing your men seem to have in hand is their hands all over the evidence. I'll give you this, your guys sure as shit know how to muck up a crime scene."

"I assure you we have our top—"

"What you've got is one guy leaning on a car that hasn't been fingerprinted—with his briefcase on the trunk just for good measure. You've thrown household sheets over the bodies and the evidence with those bodies—I happen to recognize that sheet there as one my wife bought—and the guy over there's opening that bedroom window. Why? A little warm, is he? Now, I'd like to go into my daughter's house before your *top* men manage to fuck up the entire scene."

"I'm not going to let you do that. I will call you the second I have any information."

As though Helder hadn't spoken, I continued to the end of the driveway and a small gate that opened to the front yard. Next to the gate was a wishing well—if only they worked. I placed the cigar between my teeth, giving my unsteady jaw something to grasp, then stepped onto the flagstone walk that curved to the front door. Ten feet ahead, a mass of blood darkened the sidewalk. Helder caught up. "You're not going in there, Mr. Tate."

I shaded my eyes from the relentless August sun, and looked toward the open front door. In the entry hall, two men wrestled a body duffel onto a gurney. "Is that Sharon?"

"No. I told you, she's already gone."

"Who is it?"

"Mr. Sebring."

"I want to see him."

"No—"

"God dammit, he was like a son to me."

"Then believe me, you don't want to see him like this."

One of the men closed the front door. Below the windowpane, in smeared red letters, was the word PIG. I looked away, but it was impossible to escape the implications of violence; blood was all over the porch, the grass, even the bushes. What kind of madness lay within the walls of that house?

Like toppling dominoes, the muscles tightened throughout my body. For hours, my emotions had been in constant flux, from speculation and sadness to the rage that now seared my stomach. I wiped at the corners of my mouth, swallowing hard to keep the bile down. "Tell me what happened in there."

"I don't have any information at this time. I'll be in touch tomorrow," Helder dodged.

He had the smarts of a cockroach if he thought he could out-tangle me. "Why don't we cut through the bullshit, Lieutenant. I've done enough investigations to know that everyone is a suspect to you, including myself. That in mind, I'm waiving all rights and giving you my formal statement: I did not kill my daughter. I do not know who killed her. I do not know her social activities over the past week because I have not spoken with her since the first day of this month. I can tell you this, I plan to use every means available to hunt down her killer, and I will not rest until he is behind bars—or dead. Now, you can keep me from going in that house, but I am not leaving until you tell me what in the hell happened last night."

Helder pinched the bridge of his nose as if slapped with a headache. "Okay," he relented, "but over there." He pointed at the distant lawn table and chairs.

The detective took the lead across the lawn, tracing the hedges that bordered the front of the house. Though denied physical entry, the large French windows provided inside access.

The living room looked as if a tornado had redecorated it in order to erase memories of placid evenings by the fire. Blood splashes stained the walls, furniture, and carpet. In the center of the room, a rope hung from ceiling beam to floor. Laughter escaped through the window—just another day on the job for those boys.

In the bedroom that Gibbie and Woytek shared, two officers picked through their belongings; another snapped photos of their findings.

From muffled voices slipping through the open window, one word came through clearly: *drugs*.

Helder nodded toward the same window that held my attention. "What do you know about Folger and Frykowski?"

"Not much. He's a friend of Roman's from Poland, and she's part of the coffee family. They were staying with Sharon until Roman came home. Why are they talking about drugs in there?"

Helder cleared his throat. "We found a number of narcotics on the premises: cocaine, marijuana, and various capsules."

Guilt twisted around my spine. Two weeks ago, Sharon had complained about the couple's penchant for entertaining at all hours. With the imminent birth of the baby, she needed a quiet setting. Adding further stress, she was uncomfortable with Woytek's friends, suspecting some of them were drug dealers. I told her to kick them out, but she didn't want to hurt their feelings. Had Sharon been right in her suspicion? I'd witnessed the blame game so often that I should have been able to swipe it away. But I volleyed the first pitch anyway; if I'd taken control of the situation and tossed Frykowski on his ass, Sharon would be alive.

Spread across the grass lay two white though bloodied sheets, marking the space and time of what had transpired. "Where did they find Sharon?" I asked.

"The living room," Helder said, as he pulled chairs from the table.

We sat between the pool and an outside door that opened into the master bedroom. Inside, the room was sparsely furnished. Nightstands stood on either side of the bed. A television and armoire were against the opposite wall.

Detectives ransacked the armoire drawers that held Sharon's possessions, bagging some for evidence, carelessly casting aside others that were insignificant to them; reminding me that this is no longer her home. It's her crime scene. An officer reached atop the armoire to pull down a bassinet. High above his head, he lost his grip. The tiny bed toppled, spilling baby toys from within.

I rose from the chair, intent on delivering a jawbreaker to that cop. Then Helder veered my attention. "We have a suspect in custody."

I eased back down. "Who?"

"William Garretson, the teenager living in the guesthouse. Can you tell me anything about him?"

When Sharon and Roman rented the house from Rudolph Altobelli, Altobelli hired Garretson to look after the estate. I had met him only once. "What's his connection?"

Helder explained the morning events that led to his arrest.

When Sharon's housekeeper, Winnie Chapman, arrived for work, she noticed slack phone wires draped over the gate. Preoccupied with the downed wires, she hurried through the back kitchen door and directly to the telephone. The receiver was silent.

In fact, the entire house was oddly still; so much so, that a slight movement under the table startled her. She peered below to find Sharon's Yorkshire terrier cowering under a chair. In an unusual act, the puppy recoiled when she reached to pet her.

From the kitchen, Winnie stepped into the dining room. "Mrs. Polanski? Hello. Anybody here?" Instinctively, she paused. The atmosphere felt wrong. There were subtle changes. A screen missing from the open window, flowers splayed on the floor, the drone of flies.

Beyond the dining room archway was the entrance hall. The open front door creaked as the wind gently pushed at it. When she reached to close it, her focus pulled to something written on the lower outside panel, then to splotches on the ground. Before it occurred to her that it was blood, she took another step, following the red trail into the living room.

She stopped.

Frozen, her mind tried to catch up with what her eyes already knew; she was surrounded by death.

She ran from the house, her screams echoing over the canyon walls as she escaped to a neighbor's house to call the police.

Later, while searching the grounds, officers found Garretson asleep in the guesthouse. He denied any knowledge of the murders. It was inconceivable that he could be innocent. How could he have slept through the slaughter of five people? Based on that suspicion, they arrested him.

When Helder finished, I took a beat to absorb it all. It didn't add up. One kid against five? No way. On the other hand, it was unlikely that Garretson didn't know something.

I went through the scene again. There were a lot of cars on the driveway, but one was missing. "Did you all take Sharon's car away for evidence?"

Helder looked confused. "Which car?"

"The red Ferrari."

Once Helder put out the call on the missing car, it was time to go home, something I'd been dreading since the LAPD called. The scene I was about to leave was almost as unbearable as the one I'd face at home.

Patti

Our solitude lasted for about an hour. Once the police released Sharon's name, the phone rang nonstop. While I watched over Mom, Debbie answered the calls. Although waiting to hear from Dad, the press became so persistent, she took the receiver off the hook.

Soon enough, the television vans invaded our subdivision. Neighbors gathered on the street curiously watching the press members knock on our door or peer through the windows. In a matter of hours, the media transformed us from an average middle-class family to the hottest news story on the airwaves.

We huddled in Mom's bedroom.

The time that had passed since Bill Tennant's call was the loneliest of my life. The only sounds within our home were the faint voices from the street outside. Mom said Sharon was dead, but I didn't believe it. I kept hoping that Sharon would call, and all that had happened would evaporate.

I knew Dad was home before he pulled in the driveway. From the car, he yelled obscenities at the reporters. When he came through the door, I was there, looking at him with the false hope that there had been a misunderstanding. When I saw his eyes, I knew. His hand cupped my cheek. "Where's your mama, Sugar?"

I pointed toward the bedroom. He took my hand, and I followed him down the hall.

Next to Mom on the bed, he took a sighing breath before telling us that Sharon, Jay, Woytek, and Gibbie were gone. We held hands, all of us crying, except Dad as he relayed the details. By the time he finished, he'd answered the question that I'd held inside all day. We were now only a family of four.

As the daylight hours retreated, Mom remained inconsolable despite numbing sedatives.

I lay in bed, listening to my mother's sobs from behind her closed door. Sheltered by a blanket tented around me, I cried as well. Huggles rested on my belly. My heart flared with guilt—I should have given him to Sharon sooner.

To prepare for the long night that was surely ahead, I kept a flashlight in one hand and a picture of Sharon in the other.

Nighttime and the darkness it brought have always been scary. Tonight, I imagined, it would be more frightening than ever.

2

FRAGILE

I guess I kind of live in a fairy-tale world. . . . We have a good
arrangement; Roman lies, and I pretend to believe him.

—SHARON TATE

Patti

*Sharon took me to Malibu beach. While she filmed her scenes for the
movie* Don't Make Waves, *I built a sand castle made of tightly packed
walls and towers. Castle construction turned into a group effort when the
film crew helped me with the details. The special effects man dug a moat.
Tony Curtis placed a small crab in the center, and then crowned him king
of the fortress. As a final touch, the prop man placed miniature flags on the
highest towers.*

*I'd carefully chosen an area that was far enough from the water to
avoid the high tide that eventually would come. I imagined that my castle
would last forever, with people from all over the beach visiting it for years
to come. But with the high tide came an unexpected storm. From the cover
of a nearby dock, I watched the water increasingly rise until, reaching for
my castle, a crashing wave washed it away.*

Through the days and nights following Sharon's murder, I thought
a lot about that sand castle as my entire outlook changed. Life was

fragile. I'd lost more than my sister; her murder was the wave that swept away my faith. My faith in good conquering evil, and love conquering hatred. My faith that anyone could protect me. My faith in God. Why didn't He protect Sharon? Why was He ignoring my mother's pain?

Mom fluctuated between denial and panic by the hour, sometimes by the second. Frown lines burrowed from the corners of her mouth downward, reflecting her deflated spirit. When I spoke to her, she looked at me, yet through me as if I was a puff of her cigarette smoke. I overheard her confess to Dad that she wanted to die. Before the murders, it never occurred to me that any of my loved ones would die; after, I fixated on who would be next.

I believe that homes have a personality. Overnight, ours lost its cheerful character. Drawn curtains shadowed every room to keep the media from peering in. Silence replaced the usual background chatter of the television or radio, both shut down to avoid the constantly airing stories of the murders. The smell of stale, thawing dinners replaced the aroma of Mom's cooking. Family discussions around the dinner table were awkward; the past, present, and future were too painful to discuss.

For days, only the deliverymen bringing cards, telegrams, and flowers broke our seclusion. It was odd; Sharon's friends wanted to let us know they cared by sending a remembrance, yet few called, and none visited, as if murder were a disease they might catch if they got too close.

We hadn't even seen Roman since he'd returned from London. But he and his friend Victor Lownes did call to talk with Dad about Sharon's burial. They chose Wednesday, August 13, for the funeral. There was only one problem—Sharon might not be in attendance.

The medical examiner, Thomas Noguchi, had completed Sharon's autopsy days ago; nevertheless, late Tuesday afternoon, his office refused to release Sharon's remains to the mortuary. Dad spent hours speaking with one county coroner employee after another; none had an explanation for the delay. When reasoning didn't work he resorted to threats, until finally, at 5:00 P.M., Noguchi signed the release papers. The mortuary called. Sharon would be ready for viewing at 6 A.M.

P.J.

The room was a hodgepodge of manipulation with accents of warm colors, a hint of Pachelbel's *Canon in D*, sprays of golden roses, and a plush chair that seemed it could absorb the heaviest of burdens. Each of the adornments complemented the next, except for the grossly out-of-place coffin. The funeral director, stiffer than a five-star general, lifted the coffin lid, eyeballing me as if he were about to reveal a long-lost Rembrandt. What he actually revealed branded my heart.

Sharon was in her favorite blue and white flowered dress. The studio artist who had done her makeup and hair created the illusion of sleeping life. I caressed her cheek, tracing a half-inch scar that she had acquired as a six-year-old. Everyone involved in guiding her acting career wanted her to have it removed. She wouldn't hear of it. "It's part of me," she insisted, "it's who I am."

Below the scar, I felt a rough edge. When I smudged away the makeup, I uncovered a slash wound that shattered the illusion of her sleeping. Now my eyes scrutinized, searching for more signs of violence, settling on her noticeably smaller belly. My jaw tightened over Noguchi's pointless decision to separate child from mother.

Sharon's murderer stole something more precious to her than her own life; he'd denied her the breathtaking moment just after birth of seeing her baby, touching his silky skin, or smelling his newborn hair as she kissed the top of his head for the first time.

Though Sharon claimed her pregnancy was accidental, I didn't believe her. She had dreamed of starting a family even before her marriage. Later, the men ruling her professional and personal life banned the consideration. Since beginning her acting career, her life had been gradually molded into an existence of insecure complaisance—all set into motion by producer Martin Ransohoff.

1962

Martin Ransohoff was a fast-talking, gum-chewing, two-pack-a-day smoker with an adrenaline level that sent many scrambling for cover. *The Saturday Evening Post* quoted him as saying, "I have a dream where I'll discover a beautiful girl who's a nobody and turn her into a star everybody wants. I'll do it like L. B. Mayer used to, only better."

As fate would have it, that girl turned out to be Sharon. Five minutes into their first meeting he impulsively called in his assistant. "Draw up a contract. Get her mother in here. Get my lawyer. I'm going to make this girl a star."

Sharon was only nineteen, so Ransohoff convinced her to become a ward of the court, essentially giving him control of her career without interference.

A seven-year contract with Ransohoff—four of those years as options—made Sharon a commodity in which he invested thousands of dollars. She was as big a gamble as the stock market, but he stacked the odds in his favor by overseeing every aspect of her life. He filled her days with classes: acting, dancing, singing, and gymnastics. There were doctors to help her lose weight, and a variety of coaches to alter how she walked, talked, looked, and dressed. He decided who she lived with and where. He dictated even the smallest details of what she could eat and how she should eat it, what kind of car she drove, and boldly enough, whom she dated—which in most cases was no one. "I can't fart unless Marty says it's okay," Sharon joked.

As the months rolled into a year, the joke turned increasingly sad as Ransohoff chiseled away at Sharon's self-confidence to mold what he thought would be the ideal statuesque star.

Ransohoff moved Sharon into the Studio Club, an MGM-owned apartment building with female-only tenants and an 11:30 P.M. curfew. Her first Saturday night there, she joined a group of girls and headed to the Sunset Strip.

In the early sixties, nightclubs, restaurants, rock bands, movie stars, and teenagers packed the stretch of Sunset Boulevard in West

Hollywood between Doheny and Laurel Canyon. It was the up-and-coming place to see and be seen. Stripped of social barriers, everyone fit in, and everyone partied until the wee hours of the morning.

A reporter for the *Los Angeles Herald* roamed the strip, randomly interviewing this new generation of stars and wannabes. When he came across Sharon's group, he questioned them briefly, snapped a couple of photos, and was on his way.

Early Monday morning, before the alarm had a chance to go off, the phone stirred Sharon. On the other end was the curt voice of Ransohoff's secretary. "He wants you in his office, pronto."

"You doing your own publicity now, kid?" Ransohoff roared as he slammed a copy of the *Herald* on his desk.

Unnerved by his anger, Sharon reached for the paper. A small picture from Saturday night was in the entertainment section with the caption, "Are these lovelies the next generation of Hollywood bombshells?"

"I'm sorry, Marty, I didn't know—"

"Save it, baby. You listen carefully. There are a million beautiful girls out there, but ninety-nine percent of them won't make it to a movie screen because there's nothing special about them. Within a year, they'll pack their bags and crawl back to nowhere. If I so much as sniff your name near any publication, you'll be doing the same. Steer clear of the press, and make damn sure you're not photographed again unless I say so."

Bit by bit, Ransohoff eroded Sharon's identity. She was usually hidden under the guise of a black wig and pseudonym in the few small acting roles he permitted her to take. Behind his back, she cynically introduced herself as "Miss Anonymous."

She did other things behind Ransohoff's back, too, but the biggest infringement was starting a relationship with Jay Sebring.

Jay was not only the leading men's hairstylist in Hollywood, but he carried a good amount of clout with friends and clients like Henry Fonda, Paul Newman, Steve McQueen, and Frank Sinatra. At any point, he could have used one of his friendships to coerce Ransohoff out of his no-dating policy with Sharon, but for her sake, he played along.

Ransohoff either saw Jay as an obstacle, a threat, or both. When McQueen called the producer of his next film on Sharon's behalf to request that she be given a screen test, the producer—who happened to be Martin Ransohoff—wasn't amused. Ransohoff did ultimately give Sharon a screen test for *The Cincinnati Kid*, but Sharon remained convinced that he had talked the director, Norman Jewison, into giving the part to Tuesday Weld to remind Sharon of who was in control.

Seemingly, Ransohoff's response to her relationship with Jay was to sign her up for weekend classes and a seven-day workweek. Between both of their schedules, it was difficult to find time together, but when they did, they made the most of it by having dinners out, going to parties, or taking short trips to Vegas for drinks with Sammy Davis or Frank Sinatra before their shows. But Sharon would have traded it all for a single film role.

Two years after signing her contract, and with less than a dozen bit parts added to her résumé, Sharon restlessly hounded Ransohoff. "Marty, when is this going to end? All I do is go to class, eat dinner, and go to sleep. I barely even get to see Jay."

His answer was always the same, "You're not star material yet—and you shouldn't be dating a barber." That was, until the day that Sharon announced Jay had asked her to marry him.

Then he simply said, "I give it a month."

Finally, by the summer of 1965, Ransohoff cast Sharon opposite David Niven and Kim Novak in the film *13* (released as *Eye of the Devil*). The plot centered on the famous grape harvests of France, but with a twist; an ancient Wicca sect requires a blood sacrifice to save the harvest.

Sharon left for London three weeks before filming began to join rehearsals and to prepare for her role as the cult's high priestess. She spent mornings with a European dialect coach, and afternoons with technical advisors who taught her the customs of black magic.

Considering there were thirty thousand practicing witches in England, it was easy for the producers to find and hire their king and queen, Alex Sanders and his wife, Maxine Norris. During the initial rehearsal, the couple and their process intimidated Sharon, but David Niven broke the ice when he said, "I don't know why we had to hire these two [witches]—for years I worked with many of them on Hollywood soundstages!"

Intent on delivering an intricate performance, Sharon absorbed everything Sanders and Norris shared with her—until they offered to teach her to fly on a broomstick. All she had to do was undress, and then apply a special centuries-old lotion. "No thanks," she deadpanned. "I'll stick to Pan-American."

Beginning her film career in a movie with an unlucky number as its title didn't bother Sharon. It was a different story for Ransohoff, however, as he quickly found himself producing a film riddled with problems. The project went through three directors, Sidney Furie, Arthur Hiller, and Michael Anderson before Ransohoff finally brought in J. Lee Thompson to complete it.

Kim Novak was the next of the film's casualties. Except for a few awkward moments when Sharon would catch Novak staring at her, the two women got along on the set. But behind the scenes, Novak was less passive and routinely delayed the shooting schedule by arguing with Ransohoff. No one knew for sure what the problem was but as the crew eavesdropped at Novak's dressing room door they could hear Sharon's name mixed into her tirades.

Whether it was a curse or a blessing, Novak ultimately dropped out of the role after sustaining an injury in a horseback-riding accident. With eighty-five percent of the film already completed, Deborah Kerr replaced Novak, and new writers were called in to revise the script, shaving the Novak/Kerr role and enhancing Sharon's part in the process.

During the chilly winter months, the film crew returned to the Château d'Hautefort in France to reshoot the necessary scenes.

Following *13*, Ransohoff was set to produce director Roman Polanski's upcoming film *Dance of the Vampires*. Though Roman had his current girlfriend, Jill St. John, in mind for the female lead of the film, Ransohoff pressured him to cast Sharon instead. Reluctantly, Roman agreed to have dinner with Sharon to discuss the role.

Their first meeting was "a case of instant hate," as Sharon told the story. Intent on hiring St. John, Roman did everything he could to scare Sharon away—and it worked. After an hour-long dinner of listening to Roman's bluntly cruel comments about how she didn't fit the part, he walked her home. A block from her apartment, he changed tactics by making a disastrous move to kiss her. Trying to maneuver in front of her, his foot caught hers, and down they went. Untangled from his grasp, she smacked him on the back of the head and then ran to her apartment. Before turning the key, she looked back to make sure he hadn't followed. He remained right where she'd left him on the ground, laughing. Inside, and straight away, she called Ransohoff. "That's the craziest nut I ever met. I will never work with him!"

Neither Sharon nor Roman's opinion mattered much though, because Ransohoff wasn't the sort to take "no" for an answer. Forced to work together, production began on *Vampires* in February 1966 with Sharon playing Sarah, the innkeeper's daughter who was targeted by the vampires as their next victim. In addition to directing, Roman acted in the role of Alfred, one of the vampire killers who tries to save Sarah. The rumored tension between Sharon and Roman fueled the film crew's gossip tank, and they took bets on how many days she would last.

After working with Lee Thompson, who was such a gentleman, Sharon's first weeks on the set with Roman were a culture shock. His temper and impatience seemed never-ending and at times tyrannical. He acted like a petulant child as he yelled outrageous comments at Sharon and demanded take after take. Though unnerved by his behavior, Sharon pushed forward, determined to prove herself as a professional actress.

Eventually impressed by her commitment to give him the performance he wanted, Roman became less antagonistic. Within weeks, their preconceived notions of each other evaporated. As Roman warmed up to her, Sharon took notice of the complexity within his personality. He was a confident director who took charge of every aspect of the film. When acting, he countered that strength by exposing his vulnerable side to create the role of the innocent, lovesick hero Alfred.

Sharon's perception of Roman soon shifted from aversion to intriguing infatuation. When their working relationship extended into an after-hours friendship, it was his flair for living life without limitations that really seduced her.

By the time Sharon had to film her first nude scene, she completely trusted Roman as her director. Still, she arrived on set modestly bundled in a robe. When Roman saw her, he pulled her aside. "The more you try to cover up and act embarrassed, the more everyone around you will be embarrassed. You're beautiful. Be proud of your body, it's the purest thing a woman can do. Just let go, and no one will notice a thing."

He's right, Sharon thought, as she sat naked in the tub during day after day of filming. With Roman's encouragement she felt bolder and freer, the years of inhibited societal training to cover up and shut up dissolving as quickly as the soaking bubbles around her. And within those seven days she discovered that she had only one duty in life, to just be herself.

Because the recasting of Kim Novak's role required that several scenes in *13* be reshot, Sharon didn't have much of a break between the two movies. During the back-to-back filming, Jay made intermittent trips to visit her, but even so, the separation had taken its toll. Aware that he was losing Sharon, he tried to remedy the situation by dominating her, which was exactly what she didn't want. By day, Roman told Sharon how artistically wonderful she looked as they shot her nude bathtub scenes. At night, Jay needled her on the very same subject, rejecting the idea of his future wife exposing herself to the world on a movie screen. One night, he pushed her to her breaking point, and she called off their engagement.

The Italian Dolomite Mountains, known as the land of legends, the kingdom of fates, home to kings, queens, fairies, and witches, provided a welcome distraction for Sharon after the breakup. They were also the setting for the exterior filming of *Dance of the Vampires*. The cast and crew lodged in Ortisei, a village nestled in a valley where the mountain's cathedral-shaped peaks turned fiery red at sunset.

As the ancient story goes, the king of the gnomes surrounded his kingdom in the mountains with red roses. When captured by the enemy, he cast a spell that decreed: "No one shall see my roses by day or by night." But he'd forgotten to mention dusk. Ortisei locals claim that the king's omission is the reason for the brilliant sunset phenomenon.

It was in the midst of such fantastic tales and natural beauty that Sharon and Roman began their affair. From the onset, Roman told Sharon he'd never been monogamous and probably never would be. His philosophy was to "live for this moment and let tomorrow take care of itself." But there, isolated in the Dolomites with the future seemingly a millennium away, it was easy for Sharon to lay those comments aside and hold on to her idealistic view of their relationship.

While all appeared to be going well between Sharon and Roman, things were soon to erupt for them both with Ransohoff. The previous year, Roman had signed a three-picture deal and contract that gave Ransohoff final editorial control for the US release of *Dance of the Vampires*, but the producer didn't stop with a few film cuts. Despite Roman's expressed concerns, Ransohoff gave the film a complete makeover, including a new title, *Fearless Vampire Killers*. What's worse, he dubbed over Polanski's voice with that of actor David Spencer.

Upon viewing the final cut, Roman stormed out of the screening room and straight to a press interview with *Variety*. His public chastising of Ransohoff hit the newsstands the next day. "What I made was a funny, spooky fairy tale, and Ransohoff turned it into a kind of Transylvanian *Beverly Hillbillies*."

Steaming over the *Variety* interview, Ransohoff warned Roman, "You'd better shut your trap. If you want to fight us, we've got enough money to bury you."

But by this time, Roman was on the verge of signing on for his first big-budget film, *Rosemary's Baby*, and didn't care about burning past bridges. "Fuck off, Marty," he told him. Ransohoff had already advanced Roman $10,000 for a script called *Chercher la Femme*. The next day, Roman called his newly hired attorney Wally Wolf, "I don't care what it costs, get me out of my contract with him, and see what it would take to get Sharon out of hers."

While Ransohoff and Roman were at war, Sharon was still obligated to make two additional films, *Don't Make Waves* and *Valley of the Dolls*. The end results of both projects were unbearable for her. She was convinced that Ransohoff was trying to make a complete ass of her when, at the premiere of the first movie, the words "Introducing Sharon Tate" preceded tight shots of her jiggling rear end and were followed throughout the movie with an excessive number of close-ups of her breasts.

But the greater disappointment was with *Valley of the Dolls*. After having been hired as a lead alongside Patty Duke, Barbara Parkins, and Judy Garland in a film that generated as much casting buzz as *Gone With the Wind*, Sharon was ecstatic, until the first day of filming when she clashed with the difficult director, Marc Robson. From then on, she counted the days until the film would end. And through those days, her bright outlook on being a Hollywood actress began to dim. Six months after filming wrapped she saw herself in the movie and knew her lack of faith in Robson was justified.

Valley of the Dolls shared its premier aboard the maiden voyage of the ship *Princess Italia*. The cast, the director, Jacqueline Susann, and a gaggle of press set sail from Genoa, Italy, for the first leg of the twenty-eight-day trip that included several premiers in several countries.

For nearly a month during that cruise Sharon entertained reporters throughout the day and half the night. The only time she had to herself was after midnight. When everyone else was tucked away in their cabins, she'd slip on deck to savor the calm of the moonlit sky. She'd sit there until just before sunrise, gazing at the magnificent cluster of stars. Sometimes she slept, sometimes she dreamt, but most times she contemplated her life.

During those early-morning hours, she thought about the opening lines that Jacqueline Susann wrote for the book: "You've got to climb to the top of Mount Everest to reach the Valley of the Dolls. It's a brutal climb to reach that peak. . . . You stand there, waiting for the rush of exhilaration you thought you'd feel—but it doesn't come. . . . You're alone, and the feeling of loneliness is overpowering . . . it was more fun at the bottom when you started with nothing more than hope and dream of fulfillment. . . . But it's different when you reach the sum-mit. The elements have left you battered, deafened, sightless—and too weary to enjoy your victory."

Susann's words kept her up at night ever since she first read the book for her role. She knew she was a foothold away from that peak and was scared that when she reached it, she'd be lost in the *Valley*. Of course, drugs weren't her problem. To her the valley she feared was a descent into isolation, loneliness, and a life without love.

The last time Sharon was on a ship, the family had sailed home from Italy. Back then, she had been at the bottom of the mountain dreaming of the celebrated top. What she found instead on her ascent was that an acting career was harder than she'd ever imagined—not the acting part itself, but the part that required trying to survive in a man's world. A domain where a woman, no matter how talented, is still viewed as a second-class citizen. In just a short time she under-stood that if she continued acting she'd remain in a constant battle to overcome those men. And though she was a fighter, as of late, she just wasn't sure she wanted to become the kind of woman it took to survive in Hollywood.

Shortly after the *Princess Italia* docked in Los Angeles, Sharon asked Ransohoff to release her from her contract so she could retire and become a full-time wife to Roman.

Ransohoff was shocked. "Married? Sharon, he's fucking everyone in town that has a pair of tits!"

Her cheeks burned from the flush. While she was aware of

Roman's infidelities, she believed it would all be different when their careers weren't putting them on different schedules. "Oh, Marty, it's not like that at all. It's very a European lifestyle where it's done openly and naturally," she reasoned with him.

"European, shmuropean. Baby, you're throwing away your career on that little bastard. Don't you get it, kid? Polanski only cares about Polanski."

His stinging observation reduced her to tears. "You're wrong," she sobbed. "Can't you see we're in love? It doesn't matter who he's been with, he still comes home to me. He makes me feel whole and alive. He's helping me to find out who I am; no one, including you, has given me that freedom before."

In a rare moment of weakness, he pushed a box of Kleenex toward her. "Okay, okay, if you'll stop crying I'll let you go."

Sharon left Ransohoff's office with renewed hope. The fact that Roman had yet to propose to her was incidental. What was important was that it was the last time that Ransohoff would underestimate her acting abilities. Inside her car, she leaned against the seat with complete satisfaction. Her manager warned her that severing ties with Ransohoff would come with a price tag and it did; she'd ultimately have to pay him 25 percent of her earnings until his options expired. But as far as she was concerned that was a small price to pay for the freedom she was gaining. She started the car and drove toward the beach to the house she and Roman were renting; a retreat where she didn't have to be a sex symbol and she didn't have to be a movie star.

But life after that didn't calm down as much as she thought it would. Nominated for five Academy Awards, *Rosemary's Baby* was a financial and critical achievement for Roman. Sharon's fame heightened as well. With all three of her films released in 1967, the Hollywood Foreign Press took notice and nominated her as Favorite New Female Star. Their success earned them a position with the Hollywood jet-setters, and they traveled the globe to attend parties, movie premiers, film festivals, and award shows. Initially, Sharon had the time of her life—especially after so many years of constraint. But eventually, as with everything that's done excessively, her enthusiasm waned.

By the time they married in 1968, she'd been on the road for three years. With each trip wearing more and more at her endurance, she began searching for a plateau to rest on and catch her breath.

By the time her commitment to making *The House of Seven Joys* (later retitled *The Wrecking Crew*) brought her back to Los Angeles they'd been married six months, but Roman was no closer to settling down than he was on the day she met him. On the flight from London to California, she broached the subject with him: "I'm losing touch with everyone I love. No one knows where to find me. I need a home. Someplace that says who we are; furniture that we've picked out together—a mailing address, for God's sake."

Roman didn't look up from the script he was reading. "I don't want so much responsibility. I want to be able to pack at a moment's notice, and go wherever my next adventure takes me."

Suite 3F at the Chateau Marmont hotel in Hollywood was their compromise.

Of course, Roman's growing recognition enticed his friends from Europe to join him wherever he traveled and soon enough, some of them moved into the hotel, too. While Sharon worked, Roman and his entourage canvassed the city, partying through the night. In due time, the small apartment became more of a frat house than the home Sharon had envisioned.

A Hollywood secret is an oxymoron in the gossip-ridden town so naturally Sharon heard rumors that Roman took her sixteen-hour workdays as an invitation to bed other women. To most, Sharon appeared undaunted by Roman's behavior. But her closest friends knew differently. She was in a constant emotional tug-of-war. Because she knew that his Holocaust-infested childhood inhibited him from fully loving and accepting love, she had not only tried to persuade him to feel differently but she also refused to give up on him. In the midst of some of his worst behavior she would wonder if he was trying to force her do just that. By this time, however, it had come to a point where it was impossible for her not to be hurt. When Patty Duke and her husband had separated, she offered to lease her house to Sharon. Roman dismissed the idea then, but now that Sharon had reached her limit with the endless partying she had resolved to do something about it. "I can't do this anymore," she told him. "You and your friends are devouring me. I need a home so I'm moving into Patty's house whether you join me or not."

Roman not only joined Sharon, but he made a genuine effort to mend the relationship. For the next few months, Patty Duke's house became their home, along with Winnie Chapman who did the cleaning and cooking. Roman stayed closer to Sharon, spending time with her on the set while she worked, enjoying weekend getaways together, and spending quiet nights at home with friends. It was the happiest few months of her life. And in that time, Roman's sudden transformation rekindled her suppressed teenage fantasies about adult life of becoming a mother.

Admittedly, Sharon had always lived in a fairy-tale world, observing everything through a rose-tinted veil, trusting everyone, and being a bit naïve about life. At twenty-two, when she had left for England to film *13*, she was searching for a child's never-never land, blooming with the answers to all of life's questions. Along the way, she'd found Roman, who seemed to have those answers.

To Roman's credit, his influence really did help her to grow up. Through him, she learned her strengths and weaknesses. He was the realist who showed her life for the thrill ride it is. Each climb to the top is filled with trepidation about what lies ahead while the ride down is either breathtaking or treacherous; the choice lies within each person to hang on to the handrail or let go.

Like the peaks and valleys of a roller coaster, their relationship seemed to free Sharon at times and trap her at others. The prior three years had been a long journey revealing that Roman was simply another question and Neverland a place that could only be found within oneself. At twenty-five, Sharon was ready to attend to a simpler list of needs: happiness, stability, and family. Inadvertently, Roman had helped lead her right up to the threshold of this path. And she intended to follow it even if he chose not to tag along.

"From what I can tell, you conceived around December fifteenth," the doctor told Sharon. "That would put your delivery date anywhere from the last week in September, to the first week in October." The doctor shook his head, "I still don't understand how this could have happened."

She knew exactly how it had happened. She'd secretly put a hole in her diaphragm.

Sharon's thoughts orbited around the baby as she and Roman boarded a plane for London to celebrate the London premier of *Rosemary's Baby* and their first wedding anniversary. Lying in wait, beneath the euphoria was a disquiet that threatened to break the spell. Afraid of Roman's reaction, she withheld the news of her pregnancy. She glanced at him out of the corner of her eye; instinctively, her hand covered her belly. How long would it be before he noticed she was gaining weight? She had other concerns, too. What about the next film she'd agreed to do? She'd just signed on with Roman's agent and he'd gotten her a role that would make her enough money to keep her financially secure for a very long time. Should she tell him? In London, she called the only person she trusted to keep her secret. "What should I do? Roman will be furious if he finds out I kept this from him."

Jay tried to be objective. "Everyone in this town is out for themselves. Roman's told you he's not ready for a baby. Your agent and the film producers have a lot of money riding on your next picture. I wouldn't tell any of them until it's too late for an abortion. Forget that you've seen a doctor, wait another month. No one will ever know that you found out any sooner."

Though there might be future consequences to her decision, Sharon's secret empowered her, coercing her confidence back to the surface. For the first time in her life, she knew exactly what she wanted and where she was headed.

The following week, Sharon returned alone to California for dubbing on *The Wrecking Crew*. Having given up Patty Duke's house while she was in London, Sharon ended up back at the Chateau Marmont, but staying at the hotel didn't bother her this time because Roman had finally agreed to find a permanent residence in L.A.

For over a week, a real estate agent dragged Sharon all over the city looking at houses for lease; none of them interested her until they began the climb up Cielo Drive. The gate to the estate swung away from the car, welcoming them through. Sharon stood at the edge of the front lawn near the wishing well; she tossed in a penny, closed her eyes, and deeply inhaled the intoxicating smell of orange blossoms that permeated the canyon each February. "This is the one," she told the agent. "We'll take it."

IN THE TWILIGHT hour, Roman and Sharon dined by the pool of their new home. He'd poured her a glass of Champagne, but she hadn't touched it.

"Don't you like the Champagne?" he asked.

Before she lost her nerve, she blurted out, "I'm pregnant."

"You're kidding, right?"

She shook her head.

"How? You're using a diaphragm."

With a smile, she shrugged. "I don't know what happened."

"Sharon, the timing's not right. I'm not ready, and what about your movie?"

She moved onto his lap, kissing him to silence. "Shssh, everything's going to be fine," she whispered.

From February to March 1969, they enjoyed their time together, with lazy days by the pool, and evenings filled with small gatherings at the house. It was a slow process, but Roman was adapting to the idea of a baby. As if the universe had aligned a perfect storm of life and love, Sharon felt at peace. Nevertheless, on the horizon was her next film commitment, and what was supposed to be just a brief separation from Roman.

In March, when Sharon left for Italy to film *12+1 Chairs*, Roman went to a film festival in Rio de Janeiro. He had planned to join her in Italy immediately after the festival ended, but the Brazilian authorities lost his passport and he was exiled back to England, where he began preparing for his next film, *Day of the Dolphin.* Separated by borders, they were limited to spending brief weekends together whenever Sharon could get to London.

The previous films Sharon worked on had their unique difficulties; however, *12+1 Chairs* moved to the top of her list as the most miserable experience. Because of her expanding waistline, they shot the intimate scenes first. She spent week one filming the seminude scenes with a costar who belittled her at every opportunity.

Forty-degree weather and an unheated swimming pool served as the set for week two. The scene called for Sharon to dive into the pool to retrieve a chair. For eight hours, she dove into the cold water, dried her hair and clothes, then began the process all over again. The production crew forgot to order trailers and heaters. A drafty three-hundred-year-old villa served as her only protection.

Sharon spent the third week in bed with the flu and bronchitis.

Eager to see Roman again after all the delays, she was finally able to fly to London for the Easter holiday weekend. He picked her up at Heathrow Airport in a gift he'd bought her, a 1954 Rolls-Royce Silver Dawn sedan. It was a beautiful car, but in the back of her mind, she wondered if it was a gift given out of guilt. No matter how hard she'd try to wash it from her mind, the stain of Roman's ongoing affairs remained. When the holiday ended, she returned to the Grand Hotel de la Ville in Italy to shoot the rest of her scenes. With misgivings about the state of her marriage, those last few weeks away were grueling. The physically demanding scenes pitted against her dwindling energy left her fully exhausted. Each evening she crawled into bed feeling spent and lonely, with thoughts of the "valley" first hovering then slithering through the pathways of her mind. But then she'd rest her hand on her belly to find the baby, whose tiny movements from within brought her unsurpassable joy and served as a reminder of just how much she loved her life. With each passing day the baby became more active. And with each and every night she'd soothe herself and the baby to sleep with lullabies she remembered from her childhood. In those bonding weeks, with just the baby for company, Sharon could finally sympathize with her own mother's overprotectiveness.

For Roman, the cruise aboard the *Queen Elizabeth II* that he and Sharon were scheduled to take home to the States was a luxury of time. But for Sharon, well into the latter stages of her pregnancy, it was her sole travel option across the Atlantic. As the date of their departure drew nearer, Roman hinted that he might not be able to leave on time. The day before the cruise, he broke it to her. "Sharon, I can't go with you, the end of the script needs work."

"Then I'll wait and go home with you."

"You can't, they won't let you fly," he reasoned.

"I'll go see the doctor to get permission. I don't want to go home alone," she insisted.

Determined to stay with her husband, Sharon sat in a desolate reception room, waiting for the doctor to finish an emergency call. She kept her mind occupied by reading the final chapters of Thomas Hardy's *Tess of the D'Urbervilles*.

Hardy's novel follows the life phases of Tess Durbeyfield, a young woman whose trusting nature leaves her vulnerable to the cruel and often cold society of nineteenth-century England. Early in the story, Tess's father sends her to a neighboring village to convince the affluent d'Urbervilles of kinship, and ultimately, financial support.

At the d'Urberville estate, the deceptive lord of the manor, Alec, takes advantage of Tess's inexperience by stealing her virginity. Traumatized, she returns home to find that she's pregnant. Her baby, never meant to be, unexpectedly dies.

In the next chapter, Tess meets Angel Clare, who, unaware of her history, idealizes her. The two fall in love; nevertheless, Tess's tainted past keeps her from accepting Angel's initial marriage proposals. His persistence, stronger than her will, wins out. She agrees to the marriage, but only if he will hear her confession. At fate's hand, it isn't until after their wedding ceremony that she's able to reveal her secret.

The wedding night is a culmination of passionate tension that has built between the two characters. Riveted, Sharon knew the outcome, yet somehow hoped it would be different. She hardly breathed as she read Angel's lacerating response. Instead of following the love in his heart, Angel lets the societal notions of right and wrong sway him into abandoning Tess.

That winter, Tess's father dies, leaving her as the sole provider for her family. A chance meeting brings Alec d'Urberville back into her life. Alec convinces Tess that her family's future is at his mercy because Angel will never return. Though still in love with Angel, Tess reluctantly consents to be with Alec. On the eve of their agreement, an enlightened Angel returns, and pleads with Tess for another chance. But his change of heart is too late.

Enraged at Alec for twice causing her to lose the man she loves, Tess murders him, and then rushes out to find Angel.

Together at last, the lovers spend five days together before the police find them and arrest Tess. Hardy closes with Tess's execution.

When Sharon had finished the book, the last page remained open as she reflected on the tragic depth of Hardy's love story. "Mrs. Polanski, would you like a tissue?" the receptionist offered.

Sharon touched her tear-moistened cheek. "Thank you."

"I'm sorry to add to your troubles, but the doctor won't be able to see you until tomorrow afternoon."

"But that will be too late."

"I'm sorry, dear, there's nothing to be done."

Sharon walked the distance to their house. With Hardy's story fresh in her mind, she empathized with Tess. Forced to return to Los Angeles alone, Sharon worried that her husband was abandoning her, too. A fatalist at heart, Sharon contemplated why Tess came into her life now. The idea that it was to foreshadow a change in her marriage caused a shudder.

Rarely outspoken, Sharon reserved forcing her opinion to counter another's idea of right or wrong. Many confused her deference with being simpleminded; she was anything but. If the issue was important, she slyly found a passive way of conveying her thoughts.

Before leaving for the *QE2,* she wrote Roman a note: "This would make a marvelous script. It's filled with popular debating material—sexuality and society, religion, good versus evil, forgiveness, and fate. Tess will enchant you."

She left the note on top of the book, hoping that Roman's curiosity would be piqued, and that within the pages he would uncover the deeper meaning of Hardy's original title, *Too Late, Beloved.*

P.J.

The more I pondered it, the surer I was. The only thing accidental about Sharon's pregnancy was how I found out. It happened at a fundraising party for Jay's shop in San Francisco. After a couple glasses of the flowing bubbly, Jay let the news slip, then swore me to secrecy.

I stood over the casket and ran my fingertips over Sharon's forehead, wistful of her youth, when I could rouse her from sleep with the same touch. "What other secrets came between us?" I whispered. "Who did this to you?"

I don't know when the tears I'd fought so hard to suppress finally cracked the dam, but I was powerless to stop them. Separated by the wars, there had been so many milestones that I'd missed in Sharon's life. Lost time. And now, a lifetime lost, that staying in this room another two minutes, ten minutes, or even another hour couldn't make up for. At the door, I took one last glance at Sharon. Although I'd never been a spiritual man, I prayed that God would keep her safe and happy.

3

ALONE IN THE CROWD

A good amount of the pain I carried with me was [caused
by] watching my parents suffer and not being able to help
them with that loss and realizing that all I can do is wrap my
arms around them and go from day to day and tell them how
much I love them. To me, that was like hell on earth.

—PATTI TATE

Patti

All the salesclerks at Saks Fifth Avenue knew Sharon from her films. More important, they knew her as a shopaholic, and when we entered the store, they flocked around her like hungry puppies, offering up the day's sales. Sharon turned their attention toward me. It was my tenth birthday, and she wanted to buy me a dinner dress.

Three jammed clothing racks waited for us in a changing area that was bigger than my bedroom. Dress after dress went on and came off before Sharon grinned with approval. The creamy, rich, green velvet fabric feathered against my skin as she circled, scrutinizing every inch of the sleeveless dress. "The hemline will need to be brought up an inch," she told the salesgirl. "And we'll need to find a sash for the waist."

While the clerk went to search for accessories, Sharon came up behind me, framing us in the mirror. "You look absolutely glorious."

I rolled my eyes, avoiding our reflection. She stepped in front of me. Tilting my face up, she asked, "Why do you do that?"

I smiled weakly and shrugged.

Her hands cupped my cheeks. "Inside and out, you're the most beautiful ten-year-old girl I know. Promise me that whatever happens in your life, you'll never let anything take that away from you."

"Okay, okay, I'm beautiful," I giggled.

Her expression shifted. When that somber look overcame her eyes, I knew to pay heed. "Promise me, Patti."

My sarcasm fell away. "I promise."

She pulled me in for a bear hug. "Okay, on to more serious matters; what kind of ice cream do you want?"

THE MORNING OF Sharon's funeral, I put on my green velvet dress before going out to face a house filled with relatives. Their company and the return of flowing conversation proved that life—good or bad—would continue.

Nannie Tate kept me busy helping her to create makeshift chapel veils, but she couldn't fill the emotional void I felt from my parents' absence. They both had isolated themselves. Dad left for the mortuary at dawn, and Mom had yet to come out of her bedroom. I didn't see either of them until it was time to leave for the funeral.

As the limousine coasted toward our destination, Dad explained that the media would be waiting for us at Holy Cross; all I needed to do was keep close and ignore them. What he didn't tell us was that he feared Sharon's killer would target someone else in our family. Earlier that morning he'd gone to Holy Cross with five of his friends from the FBI to set up a protection perimeter for our arrival. Along with providing protection, their other purpose was to look for suspects.

Despite his warning, I was unprepared for the hordes of paparazzi that descended on us at the church. We waited inside the car while Dad's friends pushed the crowd back to a safe distance before risking our exit.

Outside the safety of the limousine, the popping flashbulbs and intrusive questions were staggering, even with the six-foot perimeter in place. We ignored it all the best we could while we met up with Roman at his limousine. Inside the car, sunglasses hid his bloodshot and darkly swollen eyes. Mom climbed into the backseat, and I followed closely behind. Wrapping my arms around him, I buried my face in his jacket and let his lapel soak up my tears. Clinging to him made me feel closer to Sharon, and I never wanted to let go. After a moment, he gently pried my arms loose. He smiled tenderly and then broke down.

Mom had avoided Roman since his return to Los Angeles—it was just one more way she could deny Sharon's death. Now forced to accept my sister's absence alongside him, she eased Roman's head onto her shoulder and silently wept with him.

We waited in his car until the pallbearers were ready to take Sharon's casket into the church. Combined with the stewing heat, the blinding camera lights and flashes were dizzying. I gripped the back of Dad's jacket while we made the slow procession behind the casket. Nausea threatened as one unfamiliar face after another swirled, peering down at me with curiosity. I shut my eyes to slow the spin. Just as my legs wobbled, fresh air rushed through me. We were in the chapel; the doors barricaded the chaos behind us.

One door to the unknown closed while another opened to my first funeral and an intimidating crowd. A service usher led us to the grieving room to wait with friends and family for the service to begin. My parents introduced Roman to our relatives. Debbie stayed close to her boyfriend. I stood alone, listening to disjointed conversations.

Whispered was speculation of who the killer might be and why they killed. Sharon's friends were scared. Some had put in new security systems, while others had bought guard dogs. John Phillips, from the musical group the Mamas and the Papas, caught my attention when he pulled back his jacket to reveal a gun. "Any one of us could be next," he commented.

Doris Duke chatted with Sharon's friends. I thought back to the other morning, when I'd hoped for her well-being. Maybe if I'd hoped for Sharon's safety instead of hers, we'd be at a different funeral. It was a shameful secret, but I wished Doris had been the one to die.

Peter Sellers steered Mom into a removed corner. His mother had recently passed away, and he was trying to console Mom with his experience. Peter was a gentle and kind man. Whatever he shared with her was working; for the first time in days, Mom looked calm. They stayed together until Father O'Reilly came in to ask if we were ready to begin the service.

Separated into four seating areas, the sanctuary formed a cross, centered around Sharon's casket. As we moved down the aisle, many reached out to share their condolences. With my eyes focused on the silver coffin before me, I barely noticed any of them. It was so hard for me to believe Sharon was inside, that I would never see her again, never get to say good-bye. I wanted so badly to go back to the last time I saw her.

With less than two months before the baby's arrival, Sharon worked hard to finish the nursery and she'd asked if she could borrow the rocking chair that Mom had rocked all us girls in. It was only eleven o'clock in the morning, but the heat rose above eighty degrees while Mom and I carried the rocking chair to the porch. "Sharon! We're here," Mom said once we were inside.

"Be right there," Sharon called out from her bedroom.

"Well hurry up—I can't stay long," Mom said to her. Then to me, "Let's take it to the nursery."

We turned right and went through the dining room when Sharon appeared. "Mom, the room's not finished. Just put it in the living room for now."

We backtracked through the entry and set it down where she had asked, then I plopped into the chair, exhausted, while Sharon opened a giant book filled with different materials. "Come look at the swatches I've picked out for the nursery."

Mom sat on the couch that had an American flag draped over the back; she pointed at it, "I thought you were going to get rid of this."

"I am, I just don't know how to do it without hurting Gibbie and Woytek. It's their way of making a statement."

"A statement of disrespect. You'd better get rid of it before Daddy visits again because he will—"

"Shhhh. They're coming." A second later Gibbie and Woytek came through the front door. At the same time, the phone rang. *"That'll be Roman,"* she said, running for the bedroom.

"Sharon, wait, we have to go," Mom called after her.

"I'll be right back."

After a short conversation with Gibbie and Woytek, Mom looked at her watch. "Patti, let's go. I'm going to be late for my lunch."

"Wait. I want to say good-bye to Sharon."

"She's on long distance with Roman. It'll be a while. We'll be back tomorrow. Let's go," she said, and nudged me toward the door.

We didn't make it back to Sharon's house the next day. Mom had always taught me to pick and choose my battles. For the rest of my life, I'd regret not defying her in that moment.

It didn't seem fair that my parents had decided on a closed casket. Although I was forbidden, I wanted to get the coffin open. Maybe Mom didn't want to see Sharon, but I needed to kiss her good-bye.

When Father O'Reilly stepped up to the pulpit, the room hushed to his softly commanding Irish voice.

"Death violently has come among us. It has cruelly taken Sharon from our midst. There is no need for me to give even the briefest detail of the circumstances of her passing. The facts are too painful to bear even a single reference. There is one question, however, that I must ask this morning. I ask it because I believe that we must do more than mourn the passing of Sharon. Her life, her talent, the memory of her friendship calls us to transcend our present sorrow. They demand that we engage in purposeful action to wrest some meaning from a senseless deed. We can do something, I believe, and we must. We are the only ones who can answer for her before God and man.

"My question then is this: What must we, the living, do to ensure that such a terrible thing will never happen again? What must we do to bring about a world where there will be no more hate, no more cruelty, no more awful tragedy?

"We are faced with evil, but we are not without power. We can do something about this evil thing. First, we must determine never to add to the great store of wrongdoing. We can strive to create those conditions where man can be more human, more caring, much more compassionate. These were some of the qualities that Sharon exhibited in her life. She was a fine and talented person. It would be a double tragedy if we gave way at this time to despair, though our hearts are heavy.

"The world has been given to us by God, though not as something ready-made and finished. We are not robots blindly following out a predetermined plan. It is left to us to determine what good or evil will come to this world. We create in every act of good we do; we destroy in every act of evil we perform. We are, in short, free men, and being free, responsible as well. Each of us, then, must look into his own heart and see whether he has done good or evil in that part of the world that God has given him.

"In this, I see every reason for hope. We can make the world a better place. We hold the world and its future in our hands. The goodness of Sharon's life must not be allowed to pass with her. To allow this to happen would be to betray her and the memory of her, which has called us here this morning. In God's name, let us put our hands to the task and our talents to the cause of the right and the good. With that, I am a man of hope.

"To conclude, I shall paraphrase a prayer from the liturgy of the Church for the Dead. Good-bye, Sharon. May the angels take you into paradise. May the martyrs come to welcome you on your way. May we who live cherish your memory and the goodness that was yours. And may we do more. May we strive with all our might to make this world fit for men and even angels. While we are all the poorer for your passing, we are also the richer for having known you. And as with Lazarus, who was once poor, but now is rich, may you have rest everlasting. Amen."

Father O'Reilly stepped away from the pulpit. Everyone prepared to leave. Dad took my hand before leading our row toward the casket. Each of us took a rose from the family spray. One by one, we stepped up to where Sharon rested. I followed Dad's lead and placed my rose on top of the coffin. He moved ahead, until I pulled my hand from his. One way or another, I needed to say good-bye. Imagining where her face might be, I leaned over to kiss the wood. "I love you," I whispered.

I was halfway down the aisle when a wailing cry froze the procession. With her body draped across the top of the casket, Mom cried out, "This can't be the end!"

Terrified that my mom was going to die of sorrow right then and there, I raced to help her. "Mama, please don't leave me, don't go with Sharon, I need you, too," I pleaded while my arms entrapped her.

My father's grip lifted both of us from the coffin. Mom collapsed into his body, pulling me with her. When others came to help, I shrugged them off as they tried to separate us, afraid that if I did let go, I'd never see her again.

From the rear window of the limousine, I watched Dad until the car rounded the corner and he disappeared. He was staying with Roman to ride across town for another funeral.

On account of Jay's open-casket service, Dad refused to let me go, insisting it would tarnish my memory. I was devastated. As with Sharon, I wanted to say good-bye to him. Instead, a car whisked me away to Robert Evans's house, where everyone would gather after both funerals.

While producing *Rosemary's Baby*, Bob Evans developed a lasting friendship with Roman and Sharon. In a kind gesture, he made all the plans to hold the funeral reception at his home.

The drive through his estate captivated me the instant we crossed a barrier of airy eucalyptus trees. We passed gardens that were impossibly green while the rest of Los Angeles suffered the August drought. Endless beds of brilliant flowers led the way to a palace fit for any fairy-tale princess.

Competing for attention in front of the house was an ancient tree with branches that seemingly spread for miles beyond its trunk that could have easily housed Snow White and the dwarfs.

To satisfy the expected Hollywood guests, tables had been set up near the pool, and a banquet of food overflowed the tables. Music misted down from the trees while waiters circulated with drink trays. Had it been for any other occasion, this would have been a fantasyland to luxuriate in. As it was, I wanted to escape.

The sound of a television drew me into a screening room at the opposite end of the pool. Inside, the air conditioner blasted against flames blazing in the fireplace. A rerun of *Bewitched* played on the screen, coaxing me to ignore my parents' warning about watching television.

What better way to forget my troubles than to watch a show, that with a twitch of the nose anything could change. If only it was that easy.

Two months ago, we moved from San Francisco to Los Angeles. Four days ago, my biggest concern was making sure that I had cool clothes to wear at my new school. Since then, I'd discovered the harsh reality that life wasn't that simple, and I felt betrayed that my favorite TV characters had led me to believe otherwise.

A picture of Sharon filled the screen. In an instant of confusion, I thought that she had found a way to communicate with me from heaven. The next picture was of the front of her house. A man wheeled a stretcher with a body bag on it. There was a rise in the center of the bag. I tried to block out what must lay beneath it. Every instinct told me to run from the room. Unfortunately, I stayed, eager for an explanation for my sister's death.

"It's been a day of farewell to the victims of a shooting and stabbing rampage that included actress Sharon Tate and four others," a man reported. "There was talk of drugs and a mad killer, but the Los Angeles police report no developments in the search for the killer or killers that cut the telephone lines to the home in Benedict Canyon and carried out what police are describing as a methodic and ritualistic mass murder. All five victims died of multiple stab wounds, gunshots, and bludgeoning. Miss Tate and Jay Sebring were found in the living room, bound together by a white nylon rope. Tate, in her bikini bra and panties, was hung before her death."

All I had left of Sharon were memories. The images that flashed across the screen shredded them. Spattered blood covered the front porch, mounded bloodied sheets were spread on the lawn, and a dead boy was showcased in a car. Investigators ripped apart Jay's beloved sports car looking for evidence. Police officers captured Sharon and Gibbie's dogs. And hearses lined the driveway like limousines waiting to pick up the famous.

Warm memories of that house chilled to sinister ones. Why hadn't I listened to my parents? I ached to run to my mother's comfort, but I was equally hell-bent on not adding to her worries.

The sun's reflection off the pool glared at me as I bolted past a group of Sharon's friends and into the seclusion of Evans's boundless garden. A greenhouse at the opposite side looked like the perfect hiding place.

On a bench within the glass walls, I pulled my knees up to blanket my eyes. I tried to forget the news report by concentrating on the heavy scent of roses.

It couldn't have been but a minute before there was a soothing voice. "I miss her, too."

I looked up to find a man leaning against the door frame. His hair was blonder than when we had met at Sharon's house. I couldn't remember his name, but it didn't matter because I saw him as a fairy-tale prince who'd come to save me from evil.

"May I come in?" he asked.

I nodded.

He moved so gracefully that he appeared to be gliding toward me. We sat together in silence for a long time before he spoke again. "Have you talked to anyone about your feelings?"

I shook my head no.

"Do you want to talk?" he asked.

I lied with another shake of my head. There was a lot that I wanted to talk about, except I didn't know how to express my feelings.

"Fair enough," he said. "Do you mind if I talk?"

I merely nodded, but inside I was begging for someone to explain how life would move forward without Sharon.

"I know what you're thinking—you could have somehow protected Sharon if you had been with her that night. What you must understand is that there is only one thing your presence would have accomplished. Your parents would have buried two daughters today."

He was right; I did feel guilty. I had been staying with Sharon the week before and wished I had stayed longer. Whatever the outcome, I wish I had stayed. I pressed my lips together, defiantly holding back tears.

He picked up the conversation as though I were an active participant. "I've learned a trick that you can try tonight in bed. I close my eyes, and then call out Sharon's name. Within minutes, she's there, giving me that dazzling smile of hers. That's all it takes to find her."

He snapped a pink rose from the bush and handed it to me. "Now that you know my secret, I want you to remember that Sharon loved you so much that she'll never leave your side. Her companionship will be in the scent of every flower you smell, the glow in every sunset you watch, and the caress of every breeze that crosses your path. And just as she watched over you every day as you grew into a beautiful young woman, she will continue to watch over you, and guide you through all of your life experiences."

So far, I had not dared to meet his eyes. With his hand, he gently turned my face toward him. His eyes were the clearest, brightest blue I had ever seen. Warm, healing strength flowed from his hand as he whispered, "All you have to do is call her name and she'll be there. Talk to her every day, because she will need you as much as you need her."

His thoughts could not console me so soon; nevertheless, I held dearly to this brief encounter.

4

JUST FOR THE RECORD

Imagine how you would feel if people who never knew or met your daughter professed to be experts on her conduct, knew all about her, and said all sorts of degrading things about her. Much has been said about Sharon which is just not true.

—P. J. TATE

Patti

"Patti, wake up. The car's here."

I squinted against the glaring light that contrasted the darkness beyond the window. "What time is it?" I asked Sharon.

"Five o'clock. Come on, I don't want to be late. The director already hates me," she said, while dropping some clothes over my face.

The driver of the station wagon pulled away from the curb, heading west on Sunset toward Beverly Hills and the set of Valley of the Dolls. *In the backseat, I rested my head on Sharon's lap while she rehearsed her lines, "Hi Mel, the door was open. It's lovely to see you."*

I giggled.

"What's so funny?" she asked.

"You sound so different when you talk in the movies."

She tickled me, "Keep it up and I'll make you laugh until you pee your panties."

In my eyes, there were two versions of Sharon. There was her well-groomed public persona, and then there was my sister with her long, straight hair, no makeup, wearing jeans and a T-shirt.

The Sharon I identified with slouched when she sat down, bit her nails to the skin, and chewed on her lower lip when she was nervous. Her natural voice had a high timbre and a slight Texas drawl that thickened when she was excited or tired. There were days when her clumsiness got the best of her, like when she managed to break her leg getting out of bed. Balancing a checkbook was a challenge she didn't care to master. Overcoming her shyness was a constant yet well-hidden strain. She had a knack for making a mundane story interesting and loved to make people laugh. As headstrong as a person could be, once she made a decision, an army couldn't change her mind. Even so, she never forced her views on others. Instead, she'd respond, "Really? Hmmm."

On the set of Valley of the Dolls, *I sat next to Sharon in the makeup trailer as the vanity crew transformed her. They applied a heavy base makeup; first a lighter color to conceal her scars and freckles, then a darker one to emphasize the contours of her face. Black liner soon rimmed her eyes, topped by false lashes. The makeup artist brushed a deep brown color into her light brows. They pulled her natural hair back to hide it beneath a long blond fall.*

Two hours later, the finished product stood before me in a skintight gold-beaded dress with her back straight, her chin held high, and a voice that was two octaves lower with the hint of a European accent. Everything was false, from the top of her head to the lifts under her breasts, to the red nails on her fingers. It was Sharon the movie star.

WITHIN WEEKS OF Sharon's murder, another image of her emerged in the press, THE QUEEN OF THE HOLLYWOOD ORGY SCENE; A DABBLER IN SATANIC ARTS.

While the media stories surrounding Sharon's death gained momentum, I searched to find a daily rhythm in my life.

The 12:10 bell rang, signaling lunchtime. It was my third week at the new school and I hadn't made a single friend. As if my parents had made a mistake when registering me for school, my name around campus was Sharon Tate's Sister. The reaction of my classmates varied; some stared, some whispered, some looked on with pity, while others with fear. None extended an inviting smile.

As usual, I took my lunch to a table separated from the rest of the kids. I tried to keep focused on my peanut butter and jelly sandwich, but I couldn't help noticing a group of laughing girls watching me. One of them waved. "Hey, Patti, come here."

My instincts told me something was wrong, but my need for friendship won out, so I risked the encounter. When I was close enough, I saw why they were snickering. They had a magazine spread out on the table. THE SHARON TATE ORGIES: SEX, SADISM, CELEBRITIES. Along with the story were seductive publicity pictures of Sharon. On a rehearsed cue, the girls lobbed questioning stones at me.

"Are you as big a slut as Sharon?"

"Did you go to the orgies?"

"How many guys did *you* fuck?"

"Did your sister really do it with dead people and animals?"

In a burst of tears, I ran from the girls, the lunch area, and the school. I kept running, I didn't know where—it didn't matter, because I was running from life, and I never wanted to stop. I hated them all for trying to turn Sharon into a monster.

In the aftermath, Mom found me a block from home, sitting on the curb. I'd stopped running only because my lungs screamed in protest, but the stream of tears ran on. She knelt beside me. "I've been worried sick about you."

"I'm sorry. Sometimes it just hurts so bad that I can't stand it."

In the past, Mom always had the answers to bandage my wounds. This time, all she had to offer was a comforting hug.

Doris

Since the police couldn't solve Sharon's murder in a timely fashion, others tried to. Speculation about the motive appeared in newspapers and magazines with bold titles: WHY SHARON HAD TO DIE; BLACK MAGIC RITUALS; DRUG DEAL GONE BAD; VOODOO GOONA GOONA WENT BERSERK; WILD PARTIES THAT LED TO THE MASSACRE; LIVE FREAKY, DIE FREAKY.

Joe Hyams authored an article entitled WHY SHARON HAD TO DIE. Though he'd only met Sharon once, he gave readers the misconception that he was a good friend, and then hawked stories to anyone willing to pay; the higher the amount doled out to him, the more outlandish his tales became.

In one instance, Hyams said, "Sharon and Roman were in the habit of picking up strangers on the Sunset Strip and bringing them home for orgies. The way they lived, I saw the murders coming." In another story, he insisted Sharon told him she'd taken more than sixty-seven LSD trips, and "used marijuana liberally, even when she tossed her salads." Hyams also claimed that during many intimate conversations with Sharon, they'd talked about sex in the movies. Falsely quoting her, he wrote, "Sharon told me, I wouldn't mind it (nudity) one bit. In fact, I'd enjoy shocking the hell out of those Midwestern audiences." He underscored his conclusion, "She shocked them all right."

Sharon's close friends refused comment to the media; everyone else had a story to tell and the rumors rolled through Los Angeles with a relentless force.

Alex Sanders, the self-proclaimed king of the witches, insisted he initiated Sharon into witchcraft while working with her on *Eye of the Devil*, offering publicity stills of Sharon in a pentagram as his proof.

Popular astrologer Jim Slaten announced, "Sharon Tate was too involved with animal sacrifice, and they were afraid that she was going to draw too much attention to them."

Adamantly convincing when he sold his story, Slaten was inconclusive about who "they" were when the police questioned him for specifics.

Truman Capote appeared on *The Tonight Show* to share his theory. "Oh golly, it's sort of a fantasy. Of the five, I happen to know four. My feeling is it was committed by one person . . . and this person left an hour before the murders. . . . Something there triggered an instant paranoia and hatred in him. He goes home and comes back with a knife and gun.

"He rings the bell and walks inside. He ties Sharon and Jay together so that if they moved either way they'd strangle themselves. At this point Frykowski and Abigail Folger decide to run for it. He kills them. He then proceeds to kill Sharon and the hair boy (Jay) in a systematic rage. As if after an intense sexual experience, I think he went someplace and slept soundly for two days."

Capote didn't expound on what upset the killer, nor did he explain how anyone in a rage kills systematically.

Actor Dennis Hopper said that filmed orgies and whip-tying rapes were the order of the day at Sharon's house. The *Los Angeles Free Press* quoted Hopper: "They had fallen into orgies, sadism and masochism and bestiality, and they recorded it all on videotape." Later in the interview, Hopper revealed, "The police told me that three days before they were killed, twenty-five people were invited to that house for a mass whipping of a dealer from the Sunset Strip who'd given them bad dope."

If the police had knowledge of those events, they didn't write it in their investigative reports. Perhaps they had a better understanding that a woman who's eight months pregnant has limited stamina for orgies and whippings.

The country speculated on who'd been the initial target for the murders. Some thought it was the work of the mafia or the Polish underground. Others felt it had to be a drug burn. And still others insisted the murders were the result of the macabre films that Roman made. Everyone had their theory; after all, decent, innocent people weren't murdered in their own homes—not without good reason.

Newsweek reported, "Nearly as enchanting as the mystery, was the glimpse the murderers yielded into the Hollywood subculture in which the cast of characters played. Gossip about the case is of drugs, mysticism, and offbeat sex—and for once there may be more truth than fantasy in the flashy talk of the town.

"Some suspect that the group was amusing itself with some

sort of black magic rites as well as drugs that night, and they mention a Jamaican, hip to voodoo, who had recently been brought into Frykowski's drug operation. Indeed, there is speculation that the murders resulted from a ritual mock execution that got out of hand in the glare of hallucinogens."

In a *Life* magazine story, Barry Farrell's headline read: IN HOLLY-WOOD THE DEAD KEEP RIGHT ON DYING and the column continued with the words, "You wouldn't believe how weird these people were. . . . If you live like that, what do you expect. . . . Their circle may have been friendly enough to protect them in their lifetimes, but now, in their posthumous notoriety, rumor had revealed them to all as connoisseurs of depravity, figures torn from a life that was pure de Sade, with videotape machines in the bedrooms."

In two other articles, Dial Torgerson, a writer for the *Los Angeles Times*, audaciously summarized Gibbie and Woytek's lives in a couple of paragraphs. "A defector from Poland, [Frykowski] mingled with the Parisian underworld then came here and, some said, dabbled in illicit narcotics trade. He drank chilled vodka by the water tumbler, drove too fast, smoked marijuana, and took mescaline.

"Gibbie Folger came to Los Angeles to do social work. But according to an informant being guarded by the police, she became the mistress of Frykowski, financed his drug habit, and became increasingly fascinated with his study of black magic."

Had Mr. Torgerson inquired further than the rumor mill, he would have found that Woytek's study of black magic consisted of a seldom-used Ouija board that the police discovered in his closet. The board was alongside Monopoly, Tripoley, and other games.

The media's characterization of Jay made me cry, and I'm sure his parents felt the same way with each blurb they read. The implication that he was merely a sexually perverted, drug-addicted, bisexual candy man to the stars hardly described the gentle, compassionate, and generous man I knew and loved.

In almost everything documented during those few months, I was hard pressed to discover more than five accurate labels used to describe the victims: actress, hairstylist, heiress, teenage boy, and Polish émigré.

Roman thought he could end the gossip by giving Thomas Thompson, a reporter from *Life* magazine, an exclusive, in-depth interview. I will never understand why he chose to use Sharon's murder scene at the Cielo Drive house as his backdrop for that interview, but when Roman, Thompson, and a photographer arrived at Cielo, they found "psychic detective" Peter Hurkos waiting by the gate. Hurkos, best known for mistakenly identifying an innocent man as the Boston Strangler, conned his way through the gate with them.

Not surprisingly, Roman's "Tragic Trip to the House on the Hill," as *Life* titled the piece, backfired, spreading a new band of allegations. The day after publication, gossip columnist Rona Barrett led the way by reporting on KTTV, Los Angeles, that *Life* paid Roman $50,000 for the interview.

Adding further injury, Hurkos sold his story, along with pictures he'd stolen from the *Life* photographer. Hurkos began his wild tale with a brief description of his abilities as a psychic. "I see pictures in my mind like a television screen. When I touch something I can then get a vision."

The vision he received at Sharon's house: three men, friends of Sharon's, went to her house for a black magic ritual called goona-goona. They all took massive doses of LSD during the ritual, causing the three men to go berserk and turn into homicidal maniacs.

It sounded so scary. Undoubtedly, Hurkos's television antennae lost reception the day he visited Cielo. *Goona-goona* is simply the Balinese term for magic. It is also closely associated with peace, harmony, trust, understanding, and communication within Wicca society.

EASILY LURED INTO the sensationalistic stories of Sharon and Roman's lifestyle, Rudolph Altobelli, the owner of the Cielo house, picked up the press's epitaph for the victims—"Live freaky, die freaky"—and decided that the victims were responsible for their own deaths. Furious about the damage done to his home during the murders, he sent us an enormous repair bill, insisting that blood had completely ruined his carpets, furniture, and draperies—and to boot, someone tore through the front bushes, destroying his immaculate landscaping.

Upon receiving Altobelli's bill, P.J. wrote him a letter: "I can't personally speak for the others, but we taught Sharon better manners than to leave a mess for someone else to clean up. I will duly reprimand her for such careless and inconsiderate bloodletting."

The letter didn't go over well with Altobelli. He filed a lawsuit against Sharon's estate, contending that Sharon and Roman broke their lease agreement by allowing Gibbie and Woytek to live there. The lawsuit stated: "As a direct and proximate cause of trespassory conduct of Abigail Folger and Woytek Frykowski, and each of them, the Plaintiff's furniture, furnishings, and other personal property in said main house were damaged, destroyed, and depreciated in an amount not less than $15,000. . . . As a further direct cause of said conduct of defendants . . . the interior and exterior structure of said main house and adjacent lawns, gardens, and shrubbery have been damaged, destroyed, and depreciated of not less than $15,000. . . . Plaintiff is informed, believes, and alleges that as a direct and proximate cause of said conduct of defendants . . . the real property thereon have been damaged and depreciated . . . [to] an amount not less than $150,000."

After *Life* published Roman's story, Altobelli used the article as justification to sue Sharon's estate for an additional $300,000 for "embarrassment, humiliation, emotional, and mental distress." Furthermore, Altobelli noted, he was entitled to exemplary damages of $200,000 because of the "defendants' willful, deliberate, and malicious" conduct.

Oddly, in the midst of the pending lawsuit, Altobelli contradicted his ill feelings when, during an interview, he claimed he'd grown fond of Sharon. "She had a kind of warmth, a niceness, which you sensed immediately on meeting her."

Maybe in Altobelli's circle $480,000 was an attainable sum of money, but Sharon made a quarter of that in her entire lifetime, and it would take P.J. and me the next twenty years combined to come up with that money. So we spent the next two years concerned that the lawsuit would financially devastate us.

The $4,350 judgment Altobelli eventually received hardly seemed worth all the hardship. After his victory, Altobelli commented, "This was not a personal vendetta against Sharon Tate or her family, it was just business."

I suppose that during the intervening months before the police apprehended the killers, all that transpired in the media was just that, business. In any event, the price paid was too high. As the victims lay defenseless in their graves, the press and others stole their innocence, identity, and dignity as surely as the killers had stolen their lives.

5

A MILLION TO ONE

My finest, happiest memory of Sharon?
I cannot allow myself that privilege.

—DORIS TATE

Doris

It was the summer of 1961 in Verona, Italy. In a month, P.J. would complete his two-year assignment at Camp Passalacqua, and we could return to the United States.

Up until Sharon's high school graduation that spring, we were right on course, living as an average family with average problems; we had uneventful days, good times and bad, arguments and laughter. However, during her graduation dinner, Sharon veered us right off that course and into unknown territory when she announced that she wasn't going to college.

Sharon was too young to remember, but her career began when she was six months old. In 1943 P.J.'s mother submitted Sharon's pictures to toddler beauty contests. I didn't think much of Nannie Tate's efforts until she proudly presented me with Sharon's first winning title, *Miss Tiny Tot of Dallas*, and a $50 savings bond.

Through Sharon's adolescence and early teens, she entered other contests in a variety of cities as the Army transferred P.J. from one state to another. By sixteen, Sharon considered a career in modeling after she won the Miss Richland beauty pageant. Her next step would have been competing for the Miss America title, but as Sharon made plans to enter the pageant, P.J. received his orders for Italy.

When I told Sharon we were moving, she flatly announced, "I'm not going."

I laughed. "Don't be silly, of course you're going."

In an effort to stay behind in the States, Sharon pulled every antic she could think of, including the silent treatment, a starvation strike, and refusal to pack. It turned into a battle of the wits between us, but I held my own, and Sharon's refusal to pack left her scrambling through the garbage. "That's okay, darlin'," I told her. "I'll pack for you." In lieu of suitcases, I used the trash cans.

Sharon could throw a tantrum with the best of them, and the day we left for Italy she got onto the transatlantic ship all but kicking and screaming.

The rest of the family managed the transfer, but the move was tough on Sharon. It took months before she picked up the pieces by making new friends, and halfheartedly modeling for army publications such as *Stars and Stripes* and the *Army Times*. By spring we'd all settled into a comfortable routine. Then Pat Boone arrived in Italy to film his summer special.

The Venice backdrop was the perfect setting for Boone to sing his love songs. The sponsors, however, were concerned that the exotic look of European women would tarnish Boone's wholesome appeal, so they insisted that American women be cast to surround him on the show.

Boone's talent scout found Camp Passalacqua and Sharon. At first, P.J. and I were firmly against Sharon doing the show, but with only a day left to cast girls, the scout was persistent. "Mrs. Tate, Pat is rehearsing in Venice today. Why don't you come meet him and see exactly what we'll be doing with Sharon," he suggested.

Forty-five minutes later, Sharon and I arrived at the romantically picturesque Grand Canal near St. Mark's Square. When we were introduced to Pat Boone, he lived up to his charming reputation. "Mrs. Tate, if we have to, my wife and I will personally keep an eye on Sharon. You have my word, she'll be safe," he promised.

The next day, Sharon came out of the costume trailer all dolled up. A tailored blue satin dress with a matching wide-brimmed hat made her the prettiest girl on the set—until a gusting wind ripped away the hat, and wildly tossed her hair into a gnarled mess.

The shoot was equally disastrous. The gale force pushed the canal waves over the boat's edges, soaking the occupants, and made Sharon seasick for days after. God knows why, but the experience of the Boone show hooked Sharon on the entertainment industry.

Against my instincts as to what was best for her, I allowed her to work as an extra on the movies filming around Verona. The pay wasn't very much, $1.60 per day, plus a sandwich and a glass of wine at lunch. Even so, Sharon loved it. For the next year she took every opportunity that came her way, including work on the films *Barabbas*, *Vengeance of the Three Musketeers*, and *Hemingway's Adventures of a Young Man*.

On the set of *Hemingway's Adventures of a Young Man*, Sharon met actor Richard Beymer, and following a couple of dates, the two became inseparable while he was in Verona. She genuinely cared for Richard, but I imagine she also saw him as her way out of Italy and into Hollywood.

Handsome and charismatic, Richard won P.J.'s confidence and mine—at least until we found out he was prompting Sharon to go to Hollywood. That evening, during Sharon's graduation dinner, the dining room turned into a sparring arena as she grappled for independence. With all the confidence she could muster, Sharon said, "Richard said he'll introduce me to his agent in California."

"Maybe after college," P.J. said absently.

"I'm not going to college; I'm going to California," she timidly announced.

That caught his attention. "The hell you are."

"College isn't going to help me become a better actress," she challenged.

All of us jumped when P.J. slammed his hand down on the table. "Dammit, Sharon Marie, you are going to get a proper education. End of subject."

Stubborn as he was—like an irritating hemorrhoid—I seldom bothered to challenge one of P.J.'s flare-ups, and Sharon knew better, too. Beneath the table, I crossed my fingers that she'd follow my quiet example; nevertheless, she called upon her father's genes. "You can't stop me this time," she said defiantly.

P.J.'s eyes were ablaze. "Watch me. How do you think you're going to get there? And where do you think you're going to live?"

"I'll use my graduation money and savings bond, and Richard said I could stay with him until I get on my feet."

The force with which P.J. stood and toppled his chair startled us all. "Oh, you can live with Richard—as his nurse, because I'm going to break every goddamn bone in his body!"

Then Sharon shot up, too. "You're impossible!" she shouted as she tossed her napkin on the table and then ran to her room.

I let out a sigh of relief. Round one was over.

Rounds two, three, and four didn't go much better, but then the odds tipped in Sharon's favor. P.J. received his next assignment. Wouldn't you know it, we were going to San Pedro, California, just south of Los Angeles.

The news of the transfer led to a compromise. Sharon could go to California two months ahead of the family under the condition that she attend college while pursuing an acting career.

I was a bundle of nerves as her plane glided down the runway and lifted toward the States. Frightened as I was of flying, a feeling of dread overcame me as her plane disappeared into the horizon. It was our first separation.

In the early years of Sharon's life, she and I were mostly on our own while P.J. fought in the war. Trying to raise a child alone in the mid-1940s was tough enough; trying to raise a child prone to unexplained illnesses was even tougher. On several occasions, I rushed Sharon to the hospital for high fevers and spasms. Frustrated with the doctors who couldn't define her condition, I left the hospital the third time and never bothered to return, deciding I was the only one capable of nursing her. During the restless, lonely nights of willing away fevers with cold compresses, I became increasingly overprotective of her; she was all I had.

From across the sea, Sharon called daily to update me on her progress in California. In spite of our steady contact, I felt fragmented without her, and a mountain of anxiety was building inside me.

Hindsight makes everything clearer, but initially, no one noticed the clues to my impending breakdown: I couldn't eat or sleep, I did my household chores erratically, and I was listless much of the time.

Secretly, I took the train to Venice to see if a psychologist could help me through the anxiety. "Sharon's so far away," I explained to the doctor, "I'm worried she'll be hurt, and I won't be able to get to her fast enough to help."

"Mrs. Tate, what do you think is going to happen to your daughter?" the doctor asked.

"I'm not sure. I just have a horrible feeling she will be attacked or maybe even murdered."

With a dismissive wave of his hand, he smiled. "Now what do you think the chances of your daughter being murdered are?"

"I don't know."

"Try a million to one," he reassured. "Now you give that statistic some thought, and I think you'll see how irrational your fears are."

The doctor sent me on my way with a prescription for sleeping pills and a statistic, neither of which helped. Within a week, I was hiding behind so many pills, I couldn't get out of bed.

Along with the next doctor came a diagnosis of acute separation anxiety disorder. The doctor offered only one quick remedy. Sharon unhappily returned to Italy.

Eight years later in California, Sharon's death raised all the same symptoms, the listlessness, the depression, and the crying spells. The attending physician gave virtually the same diagnosis, only this time he didn't have an immediate solution.

Some referred to my illness in Italy as a nervous breakdown. After Sharon's murder, I referred to it as a premonition. There was another premonition haunting me; a dream I had just after Sharon moved into the Cielo house.

In the dream, I'm with Sharon in the Cielo living room. An uninvited cowboy comes through the front door. The two of us are on the couch while the cowboy stands near the fireplace, chatting up a storm. He seemed friendly enough until he ripped two guns from his belt. His voice, now a throaty baritone, avowed, "Today you will die."

The first blast of the gun sent me to my knees, crawling behind the couch for shelter from the next six shots. I assumed Sharon had followed, but when I turned, I was alone.

As suddenly as the uproar began, it ended. In the stillness, I gathered the courage to peer around the back of the couch. Only a cloud of smoke remained where the cowboy had stood. "Sharon?" I called out. From the silence, I knew she had never made it off the couch.

P.J.

Everyone deals with grief differently. Roman's way was to close and seal the chapter of his life with Sharon. On the day of the *Life* magazine interview, Roman took his personal belongings from the house, and then left behind everything else that encompassed their marriage, including Sharon's Ferrari and estate, which he signed over to me.

Plagued by Altobelli's lawsuit, I went to Cielo in order to restore the house to its premurders condition. Victor Lownes had arranged for a studio crew to clean the house, but apparently, they took one look and hightailed it back to Paramount. Driving through the gate, I wondered if I'd be any braver? On the apron of the driveway, I drove past where Steve Parent's car had been and thought about my wife's dream. It was crazy to give it any validation. Yet I couldn't help myself as I parked next to the house.

The grounds were oppressively quiet. Aside from the shadows of the past and a guard posted at the gate, I was isolated—an existence in which I was becoming increasingly adept.

On the porch I sidestepped pools of blood, now sunbaked black. I paused at the front door. Upon each occasion I'd entered this house before, my adrenaline rushed with excitement waiting for Sharon to answer. This time, my adrenaline rushed as well as I looked at the blood inscription on the door left by the killer. The sun had faded the red letters. Still, they blasted a warning of what lay beyond. My hand shook as I reached for the doorknob, and while I did, the excessive black fingerprint dust caused me to wonder if my hand overlay the killer's imprint. The instant I opened the door, the house came to life with a broiling exhale that eased past me as though it had waited patiently for just that moment to escape. I followed the blood trail into the entry hall where I found two blue steamer trunks side by side. A piece of clothing was caught in the closed seam of one of the lids. I knelt between the trunks before unhinging the latches. Sharon's unmistakable scent escaped the confines when I opened the lid. The blood beneath my knees was momentarily forgotten as I reached for the bottle of perfume that had leaked down the side of the case.

Beyond the entry hall, the rancor of the murders remained in the living room with a morose fear that unsettled the atmosphere. Overturned pillows, books, scripts, pictures, upended plants, knickknacks, and the bloody sheets that must have covered Sharon and Jay's bodies were strewn over the floor. Some of the disarray was the result of the murders, the rest from the police looking for evidence. But the blood; the police couldn't be blamed for that. There were scarlet splashes in every direction.

The front window squawked in protest as I raised it. Sharon's bronzed baby shoe was wedged between the glass and screen that separated the contrasting scenes of life and death. No matter how big she had grown, I always thought of her as my baby who'd once fit into this shoe. I turned back to the scene inside and noticed the rocking chair that Doris used to soothe Sharon in when she was a baby. The chair had witnessed the beginning and end of her life. How could this have happened? How could her life have ended like this? Who had betrayed her? A variety of scenarios played through my mind.

The highest cause of death of women in their eighth month of pregnancy is murder; 95 percent of them are killed by their husbands. Roman's alibi lay in England, but he could have hired the assassin. Truth be told, I didn't have a clue as to what made the man tick; he spent only a small portion of his time in Los Angeles over the course of their relationship. In fact, they'd been together for months before I met him. The day I did finally meet him didn't go as well as Sharon had hoped, but first meetings with me seldom did.

I PLANNED WHAT *Sharon called one of my "surprise attacks," where without warning, I'd stop by to check up on her.*

I arrived before Roman came home from work. We sat on the deck of their rented Santa Monica beach house, where the ocean breeze wiped away the September heat, and Sharon nervously chain-smoked. I studied her face, wondering if I could still tell when she lied. "Sugar, are you happy?"

Her face beamed. "Oh, Daddy, he makes me very happy, wait until you meet him. He's sensitive yet strong, intelligent, incredibly talented." She paused, searching for the right words. "I don't know how to define him, you don't notice any one part of him, he just comes at you in one dynamic blast."

"Like a fart?" I asked, stone-faced.

She rolled her eyes. As if on cue, the thunder of Roman's Mustang came from the driveway. "He's home," Sharon said, stubbing out a cigarette. "I'm going to make a drink. Do you want one?"

I nodded.

She started away, then paused. "Daddy, give him a chance. Okay?"

I nodded again. Alone for the next five minutes, I did what came naturally; I listened and gathered information. Though hushed, I easily heard their conversation from the kitchen.

"Oh God, Sharon, I'm too tired. It's been a long day. I just want to get into a hot bath and relax," Roman complained.

"He's only staying for dinner. You need to get to know him," she said.

"Is there any grass in the house?" he asked.

"Please don't do this."

A momentary silence. Then, sounding defeated, Roman said, "Shit. Come on, let's get this over with."

They came to the deck holding hands, until he reached for mine. "I'm Roman. It's nice to meet you."

"Likewise," I said, looking him over as though evaluating an enlisted man.

The phone rang. Neither made a move to answer it. Scared to leave me unsupervised, Sharon nudged Roman. "It's probably for you."

"I don't want to talk to anyone. If it's for me, tell them I'm not here."

She hesitated then gave me a glaring look to behave. "I'll be right back."

The two of us stood there, looking uncomfortably at each other. Roman cleared his throat. "Listen, I know you're upset that Sharon and I live together—"

"It's okay, we're getting used to the idea. Welcome to the sixties and all that crap."

Roman looked one way, then the other, anywhere but at me. He forced a laugh. "You know, she's too nice. I've been trying to toughen her up, but she won't fight back."

"I wouldn't try too hard. She doesn't get mad very often, but when she does, oh, son, you better watch out." I smiled slyly, toying with him. "And when she's done with you, then you've got me to reckon with."

While my smile remained, Roman's faltered. He glanced toward the house. "I'd better go see who's on the phone."

I CLOSED MY hand around the tiny shoe, intent on bringing Sharon's killer to his knees, whoever that was.

Piece by piece, I sifted through the living room. I picked up the script for *Valley of the Dolls*, recalling the night my mother phoned after the film opened in Dallas, Texas. "Oh, P.J., I'm so proud of Sharon Marie. You should have seen the line to get in. It went clear around the block. We had to wait an hour just to get into the theater."

"Well, did you like it, Mother?" I asked.

"No, it was trash," she whispered, "but don't tell my grandbaby that. You just tell her Nannie was beaming with pride the whole night."

I picked up another script: *The Wrecking Crew*. The army base was always slow to get movies into the theater. The film was released in February but it didn't make it to the base until August. On the eighth, I'd gone to the theater in the afternoon to buy my eleven o'clock ticket to ensure I had a seat for that night.

Sharon had enjoyed working on the picture and it was apparent in every moment of her performance. I left the theater immensely proud, with a plan to call Sharon in the morning.

Based on the LAPD's interrogation of Garretson, they estimated that the murders took place close to midnight. My new friend, guilt, tapped my shoulder to remind me that while I laughed in a darkened theater, Sharon was no doubt here, fighting for her life.

For as long as I could remember, Sharon was in the habit of squirreling away every memento. Letters, cards, flowers, and everything else were stored in my old cigar boxes. The day she moved into Cielo, I was surprised to see her unpacking them. "Why don't you get rid of that stuff," I'd asked.

Sharon stared pensively at the boxes on her lap. "I just can't. If I did, I would be giving up who I am, and giving in to that," she said, nodding toward a movie magazine.

Surrounding me were those fragments that, when combined, formed Sharon's essence. As if packing for a downed soldier, I meticulously stowed each of the items in my old army footlockers, regardless of how trivial they seemed.

In the course of gathering Sharon's belongings, I had traversed her lifetime. By the time I finished, all that remained in the house to prove her existence was her death. Altobelli was right about the carpet in the living room, it would need to be replaced; however, the rest could be restored.

Flagstone covered the entry hall, porch, and walkway, and blood covered the flagstone. Without bothering to dilute it, I poured bleach directly onto the dark masses. On contact, the chemical ignited a sizzling reaction. Slow to defeat, it took several forceful strokes of the scrub brush before the crimson yielded to pink and finally to a sallow foam. Defenseless to my weakening composure, I welcomed the caustic fumes that masked the real tears I couldn't restrain.

Los Angeles is misleading. Despite the hot sun, it's a cool-climate area. As the sweltering air eased back with the receding sun, my knees were bruised, my back ached, and my hands were cramped from scrubbing, and the blood I'd cleaned branded a scarring wound I'd likely never forget. Exhausted, I roamed from one room to the next, looking for anything I might have missed.

In the kitchen, I opened the refrigerator to retrieve the last remaining item, a Heineken beer. Owing that it was Jay's favorite, I made the gesture of a toast as if he were standing before me, and then headed out to the porch.

I walked beyond the guesthouse to the edge of the property where the sky flamed with the last sliver of the sun. Life had been beautiful here for Sharon. High above the insanity of Los Angeles, the setting held the promise of a new beginning for her and Roman. But somehow the insanity had sought her out, defying the million-to-one odds.

6

NOTHING BUT DEAD ENDS

I could put forward the theory that Sharon's murder was committed by the Creatures from the Black Lagoon and the Hollywood crowd would listen and nod their heads in agreement. But at the first mention of narcotics, they all become deaf, dumb, and blind.

—P.J. TATE

Patti

On the drive from California to Texas, Dad said we were going to visit Nannie until "the dust settles at home." But I knew the real reason; my parents were as scared as I was.

Murder: the unlawful and malicious or premeditated killing of one human being by another. I added the definition to my vocabulary. The meaning was clear to me, but why someone murdered was not. In Sharon's case, the police were mystified as well.

Not long after our arrival in Texas, Dad left us behind for San Francisco to complete the paperwork for his army retirement. It was also the first leg of his investigation.

P.J.

Action quiets the mind. I could no more sit around waiting for the LAPD to solve Sharon's case than I could leave a downed soldier to die alone in the field. Five had been downed at Cielo and left in just such a manner. Their killers would pay for that heartless deed.

The previous December, after Jay opened his San Francisco salon, he leased a houseboat that served to entertain clients in the Bay Area. A fisherman at heart, I escaped to that boat every weekend. After a day of cold beers and watching the lure dip and swell, I dropped my trash at the Dumpsters. Each time, among the rubbish, I saw dozens of emptied bottles of rubbing alcohol, a necessary ingredient for making LSD.

With all the hubbub of a drug motive, I wondered if Jay had signed his own death certificate with LSD manufacturers from the harbor.

At the army base, I loaded my .45-caliber pistol as a civilian, then headed to stake out the marina.

The glow from the floodlights mounted above the dock splintered through a crack in the otherwise drawn curtains of Jay's houseboat. The lapping waves and creaking mooring lines were my only company, leaving me vulnerable to a new enemy, my imagination. If I wasn't diligent in focusing on the task at hand, the enemy crept through, graphically detailing Sharon's death. Usually, I was quick enough to hoist up the barrier. Other times, they ambushed me, threatening to flick me off my fingertips' grip on life. With a long tug of beer, I raised the wall, concentrating on the investigation.

Two weeks short of October, Lt. Helder updated me with the LAPD's progress. Every lead had fizzled out, including Cielo groundskeeper William Garretson. The case threatened to go cold.

I was luckier than I would have guessed. Car doors slammed shut in the distant parking lot, interrupting my thoughts. Three men unloaded boxes from a Cadillac Eldorado. Handicapped by the load, the men unsteadily moved across the docks to a houseboat in the last row. Through binoculars, I saw what I was looking for, a box marked ALCOHOL.

My impulse was to confront the men, but if they were drug manufacturers, my sole .45 would be futile against their probable arsenal. Opting for the safer route, I slipped out, got their license plate number, and went to my car in order to track them.

Banging car doors woke me from a fitful nap. A white blanket of fog rolled through the marina in the predawn hours. I remained slumped in the driver seat, waiting to give the Cadillac a safe lead before following.

A mile from the shore, the fog still clung heavily. My headlights were a beacon pleading for attention. At the barren Pacific Coast Highway, the Cadillac headed north while I aborted, entering the southbound ramp for Los Angeles.

I returned to an empty home, exactly the way I wanted it for the next few weeks. I dropped my duffel bag in the foyer, unbuckled the safety harness around my pistol, and then searched the house until I was comfortable that nothing had been disturbed. Along with that newly acquired habit, I checked for a connection tone before dialing LAPD robbery/homicide.

Helder was unhappy about my investigation. "I'll check the tag, Mr. Tate, but you should leave this case to my department."

"*Bob,* if you insist on formalities, I prefer 'Colonel,' but since we're going to be talking quite a bit, P.J. will do fine."

"Colonel Tate, I think your private investigation will only prove to be a negative experience for all parties concerned."

"How's your investigation going?" I chided. "Have you got enough men helping you out? I ask, because I'm confident that my network of connections can accomplish in one day what will take your men a week. Now you can accept the information I hand over, or you can let your foolish pride disregard it." Not interested in Helder's response, I hung up.

Many speculated that the word PIG left in blood at Sharon's house was a calling card from the revolutionary party the Black Panthers. The license plate of the Cadillac belonged to a high-ranking Black Panther with a long arrest history. Even more interesting, the investigators linked him to Abigail Folger's social work. But by the end of the week, the suspect's airtight alibi dashed everyone's excitement.

Helder called, using the disappointing news as a peace offering. "P.J., worst-case scenario, you helped San Marin uncover a drug operation that they've been searching six months for."

"PEOPLE THINK OF him as a monster. God, he's not a monster, he's an amazing man." On the news, Gene Gutowski defended his friend and business partner, Roman Polanski.

I turned off the television. There were three other men in the family room, all friends from the U.S. Department of Defense, all skilled investigators. After the murders, Guy was the first to offer his assistance. As an FBI agent, he'd spent twenty of his forty-nine years traveling the globe, developing an unsurpassable finesse for gathering intelligence. His tanned, handsome features and trusting smile were the façade of a shrewd mind that never rested. "Is Polanski a monster, P.J.?" Guy asked.

"Hell, I don't know," I said, pacing a canal into the carpet. "In hindsight, I don't know shit about him, except he was unfaithful to Sharon."

"He sure looked convincingly sad. He cried through that press conference a couple of weeks ago," added Frankie—short for Frankenstein, the only suitable name for a man six foot eleven.

Jake, the street-savvy one of our group, lit a cigarette, adding another cloud to the haze. "He's an actor; they turn that crap on and off like a faucet. Let's follow him around for a few days and see what turns up—besides his dick."

"Sounds good," Guy said. "In the meantime, I'll go through Interpol and see what they have on him. Frankie, you run a check on his paper trail."

Two hours after splitting off, Jake called. "P.J., did you know your son-in-law was submitting a polygraph today?"

"Really?"

"Yep. My guy inside says Polanski's hooked up as we speak. Not that it will do any good; he's probably doped up to the gills."

"You work it from your end, I'll work it from mine, let's get a copy of that tape when they're finished."

Late into the afternoon, I was still looking for suspicious entries in Sharon's check registers when Helder returned my call. "P.J., Bob Helder here—no."

I laughed. "You didn't even give me a chance to ask."

"Yeah, I know your guy was down here sniffing around for a copy of the polygraph. Polanski passed, end of story."

I didn't have the proof, but took the gamble. "Do you have any idea how much Valium Roman took before he tested?"

Thoughtful silence from Helder's end.

"Enough to tranquilize a gorilla. I know him," I lied. "I can tell you better than that machine if he was truthful."

"Dammit, P.J., what you do on your own is no skin off my nose, but if I start sharing information with you, and you act on vengeful emotions, I'm screwed."

"I won't. You have my word."

"So help me God," Helder relented, "if I find a suspect in some isolated desert grave—especially your son-in-law—I'm coming after you. Meet me at the Hamburger Hamlet at six."

DOMESTICITY WASN'T ONE of my stronger traits. Under my command, I kept the house precisely organized beneath a dusty film that comfortably settled over every surface, including the kitchen table, where beer cans cracked open as a tape deck whirred with static. Jake was the only one absent; he was on the night shift, watching Roman.

Lt. Earl Deemer submitted Roman's polygraph. His voice came through the speakers first. "Have you taken any medications?"

"A small amount of Valium at four this morning. . . . How long will this take?" Roman asked.

"If we get along all right, about an hour."

"I ask you because I want to lie one or two times, and then I will tell you after."

"Okay, if I don't tell you first. . . . Was Woytek associated with the Polish Army?"

"As far as my knowledge? Not at all. . . . He went to the university to study chemistry, Woytek was an old friend of mine. I resented him. He was a loser and whatever he started, he would fuck it up. But when he was in Paris, when he defected, he was always writing me letters, he really loved me, you know because I got him into film school and things like that. . . . But I was trying to stay away from him because he was into trouble. Then I saw him years later in New York, I saw him change very much. He was full of good spirit . . . and he'd met Gibbie Folger, a rich girl, and I thought, that's good for him. Exactly what he needs. And when he came here I was trying to help him find a job. And recently he was getting uptight about not doing nothing and I said, believe me, you have a job on my film. A few letters that I just found in London from him are full of enthusiasm. And he was reading all books about dolphins because that was the subject and ideas that he had. He really loved me." [. . .]

"Tell me about Jay."

"When I came to Los Angeles and started living here for two years, I saw more and more of him. He used to come to our parties, and I started liking the guy very, very much. He was a very sweet person. . . . I'm sure in the beginning of this relationship there was still a love for Sharon, but I think gradually it disappeared. I'm quite sure of it. And in my mind when I learned about the tragedy and the description I thought maybe, jeez, Sharon's pregnancy brought back in his mind the end of everything. But I'm positive there wasn't; there was nothing from Sharon because she was so much in love with me, as much as a human being can be."

"So, no indication on your part that Sharon went back to Sebring at any time after you two were together?"

"*No, not a chance.* Sometimes I was thinking about it, suspicious, but I'm the bad one. I always screw around. That was Sharon's big hang-up, we had *endless* discussions about it. And I said, you told me you don't want to change me. But Sharon was absolutely not interested. . . . There was not a chance of any other man getting close to her."

"Why don't you take a look at these pictures."

Roman leafed through recent photos of Jay and Sharon arm and arm. "These were probably taken by Woytek while I was in London. Sharon's very pregnant here."

"Well, what I was getting at was that he [Jay] spent an awful lot of time up there—"

"Hanging around," Roman finished the sentence. "I was on the phone with her like every two days. The only thing she talked about was, 'When are you coming?' " [. . .]

"Okay, Roman, let's get started. Keep still, with your feet flat on the floor. Take a deep breath. Good. Now let it out."

Roman interrupted. "You know, those pictures really shook me up, because I didn't see any pictures of my wife that pregnant before—not since the murders."

"Uh-huh, well, I'm going to show you some other pictures; just say yes or no if you've seen the people before."

"What if I'm not sure? I meet so many people, and I reject the faces that don't interest me. . . . I may have seen them at a party or something, but don't remember them."

"Okay, let's look through them once, then we'll go through them again with the polygraph running."

Roman only recognized one person. "That's Billy Doyle . . . I remember him because he crashed a party that we gave. He came in, and he was trouble and I said, 'Gibbie, who is that little jerk?' And she said, 'Billy.' And I said, 'Get him out of the house.' And they got him out of the house, and he came back again because apparently his car broke and I said 'I want him out,' and he was drunk, and they said, 'He's crazy, he's an idiot something's wrong with his brain,' so I remember him."

"All right, I'm going to turn this on. Stay relaxed," Deemer said, followed by a crashing sound. "Got some knives here that I'll show to you one at a time—don't hold your breath—I need you to breathe normally."

"Breathe normal," Roman mocked. "Not so easy when somebody tells you to do that."

"Just give it a try. Now, did you ever see anyone at the Cielo address with a knife such as this?"

"No," Roman commented to all ten knives, and then added, "Can I say something? I keep thinking about my breaths now."

"Yeah, I noticed."

"Only two knives looked more familiar than the others. This one. This looks a little familiar and maybe I saw someone working with a knife like this one."

"See, your breathing went all to hell here. See the difference here and then how you're breathing here? I'm going to ask you some regular questions, Roman, just answer yes or no. Do you have a valid California license?"

"Yes."

"Have you eaten lunch today?"

"No."

"Do you know who took the life of Woytek and the others?"

"No."

"Do you smoke cigarettes?"

"Yes." Roman burst out laughing.

"You know with that screwing around, I'm going to have to start over. Look at the increase in your blood pressure when you start to lie about your cigarettes. I'll tell you, if you ever do something wrong, don't take this. You're a good reactor."

"I just wanted to know if this really works. I'm sorry; I promise I won't lie anymore."

"Okay, do you drink milk?"

"No."

"Did you have anything to do with taking the life of Woytek and the others?"

"No."

"Do you feel any responsibility for the death of Woytek and the others?"

"Yes . . . that I wasn't there."

"Have you dated any airline stewardesses since Sharon's death?"

"Yes. Well, I haven't dated; I've seen a couple of them."

"Took them out to lunch or something?"

"I fucked them," Roman proudly announced.

"Okay. . . . Did Sharon fuss with narcotics, other than pot?"

"No. She did take LSD about sixteen times. At first, she liked it because it helped her to get over her inhibitions. But she'd arrived at a stage where she knew, one more trip, and her mind would be gone. So she gave it up. One of our first meetings. . . . I had a tiny amount. We split it, and then spent all night talking. . . . In the morning, she started flipping out and screaming. I was scared to death. I couldn't do anything for her. Later, she said, 'See, I told you I can't take it anymore. This is the end of it.' And it was the end of it, for her and for me. Anyway, I can assure you that she didn't use any drugs for four years, except pot, and not too much and certainly not during her pregnancy. There was not a question of it, not even a drink of wine."

"Doyle's supposedly a pretty big drug dealer. To your knowledge did he turn Jay on with cocaine?"

"Not to my knowledge. Billy Doyle's a character that Sharon wouldn't even talk to."

"Did you ever come into the bedroom and find Billy Doyle with Sharon—not in bed."

"I only saw Doyle a couple of times in my life . . . once at our party, and I saw him around Mama Cass. But with Sharon, I don't think she even knew who Doyle was."

"What's your impression of Mama Cass?" Deemer asked.

"Bad news, from beginning to the end. I hardly know her. . . . Bad news, I mean just stay away from her. . . . I didn't like her from the first time that I met her."

Roman was an experimenter on unsuspecting minds. He set Deemer up as his next conquest by pulling a cigarette from the pack on the table, lighting up, and leisurely inhaling.

"I thought you said you didn't smoke cigarettes?"

Roman didn't respond because it was a con; he wasn't a smoker.

The tape spun with a hiss of vacancy. Guy and Frankie glanced at each other, waiting for me to go ballistic over Sharon's drug use. Guy spoke first. "P.J., did you know Sharon was using?"

"No. I guess it's not so shocking. These days it seems to be the ordinary instead of the extraordinary."

Frankie slapped me on the back. "At least it's all out in the open. Now we know what we're dealing with."

"You know, he comes off just arrogant enough to be the kind of guy to knock off someone he didn't want around, 'I reject faces that displease me,' " Guy parroted Roman's accent. "What a shod."

"I don't know," I added. "Roman's tweaked, no doubt, but I don't think he's our killer."

"I'm not so confident," Guy said. "He played Deemer—gives him the lie just after the key question of the murders, plus factor in the sedatives. He might have jumped the system. Let's stay on him at least until Interpol comes through."

Ten o'clock the next morning, I urged the coffee to wake me up while I shaved. The front door slammed, then Guy's voice. "P.J.? Got the workup on Roman."

I found him in the kitchen. "Well?"

"He's a damned Communist, but that doesn't make him a killer." Guy dropped a thick file on the counter. "Nothing out of the ordinary. That includes Frankie's paper drive."

Jake and Frankie were laughing on their way up the driveway. I met them at the front door. "What's so funny?"

"I was telling him about Polanski's escapade last night," Jake said. "Hey, got any more of that joe? I'm desperate like a junkie."

I poured a round for everyone while Jake told his story.

"So there he is, middle of the night, looking like a Christmas elf, climbing over the gates of this Tudor estate—right across from the Beverly-Fucking-Hillbillies house! Only in L.A. Anyway, all he does is riffle through a Rolls-Royce near the garage. Just as he's making his escape, LAPD rolls up. They check his ID, and ten minutes later, he's on his way. So I corner the patrolie writing up the report, who tells me it's John Phillips's house. Polanski told him he left something in

Phillips's car and didn't want to disturb him—thing is, he left empty-handed."

"What's John Phillips got that he wants?" I asked.

Guy sipped his coffee. "Who knows, but let's check into Phillips and the whole group. Roman has a strong distaste for the fabulous Miss Cass Elliot."

Jake daintily held out a file. "Before my soirée in Bel Air last night, I smuggled out a copy of the detective's progress report." He shook his head in disbelief. "These guys are chasing their asses. Listen to this—and I quote: 'If we assume that the killers went there merely to conduct some type of business, such as a narcotics sale, or to enjoy a narcotics party and the killing occurred, then investigators are of the opinion that the suspects cut the telephone wires in an effort to gain as much time as possible before the crimes were discovered.' "

Jake removed his glasses. "Who the fuck did they think was gonna call out after they butchered everyone? Jesus, Joseph, and Mary?" Jake stopped short. "Sorry, P.J., but that's Homicide 101. Those lines were cut before they went in—premeditation."

I rubbed my tired eyes. "It's okay, you're right. I checked those wires myself. Out of all the ones connected to that pole, they only cut the phone and the communication wires between the house and the gate—the gate is the important one. When the button is pushed, a buzzer goes off in the house. Whoever cut them knew the lay of the land."

I opened a box of cigars and passed them around. "We need to look at those closest to Sharon. Let's check my daughter and her friends inside and out. No pussy-footing around me, I want to know everything that was happening in their lives."

AS AN INVESTIGATOR, I counted on the element of surprise to eliminate rehearsed stories. Though Jim and Wende Mitchum were good friends of Sharon's, I'd never been to their house. When Jim opened the door, I spied the nervous expression I'd expected and wanted. "Colonel? How are you? Come in. I wish we'd known you were coming, Wende's out shopping."

There's his first lie. I had spotted Wende through the window when I pulled onto the driveway.

"Can I get you a drink?" Jim offered from a stocked bar.

"No, I can't stay long. I came to see if you had any suspicions about what happened at Sharon's."

"Umm, not really, I haven't been up there in a while."

"What about the drugs?" I tested.

"None that I know of—"

"Cut the crap, Jimmy. I don't need sunshine blown up my ass. Level with me. And while you're at it, get Wende."

She shyly came around the corner. "I'm right here." She hugged me. "I'm sorry, P.J. I'm just scared."

"I know, Sugar, that's why you have to be straight with me."

"The truth." Jim sighed. "Things were pretty out of hand up there while Sharon and Roman were in Europe. Jay and Woytek had become close, and they were both doing a lot of drugs."

"What kind of drugs?"

"Jay was a coke freak. He'd been doing it casually for a couple of years, but lately it was out of control," Jim said, while he eyed Wende for support.

"I think Sharon's pregnancy put Jay over the edge," she added. "He never expected that marriage to last—hell, no one did—but I think he was waiting for them to split. With the baby, he lost hope, and buried his feelings in the cocaine."

"What about Woytek?"

"Woytek was one of those guys that had to try everything once," Jim said. "He lived every day like it was his last. I don't think there's a drug out there that he hadn't tried."

"Where were they getting the stuff?" I asked.

"I'm not sure about Jay's connection. Talk to Steve McQueen; he cleaned out Jay's house before the police got there. Woytek mixed with these nuts from Canada, Billy Doyle, Pic Dawson, and some others."

"Dawson and Doyle have been coming up a lot."

"Cass Elliot brought them into the circle. Poor Cass." Jim shook his head. "She's so easygoing, and guys like Doyle and Dawson zeroed in on that. They lived with her, spent her money, and supplied her with the hard stuff, like heroin."

"Do you suspect either of them for the murders?"

"They're both crazy," Jim said. "Rumor has it that one day Doyle freaked out on some mescaline, and then he pulled a gun on Woytek. Jay was there, too. They wrestled the gun from him, and then tied him to a tree until he came to his senses."

"What about Dawson?"

"He's an idiot," Jim laughed. "He couldn't organize a trip to the bathroom, let alone kill five people. One thing about him though, his father was a diplomat. He'd light a joint and toast, 'diplomatic immunity,' because he'd smuggled the stuff in sealed diplomatic packages."

"Was Woytek dealing?" I asked.

"Nah, man, he was the Robin Hood of dope," Jim said. "He'd get the money from Gibbie, buy the drugs, and then give them away. He only wanted to have fun. Damn shame if that's what got them all killed."

"You said things were getting out of hand. In what way?"

"Just that whole Doyle, Cass Elliot, John Phillips crowd hanging around there, and then they brought friends. It snowballed into a bad scene. A killer could be hidden in any one of those fringes."

"What about after Sharon came home?"

"She felt uneasy," Wende cut in. "Right after she returned, she had complete strangers walk into the house without knocking. She took about two days of that, and then closed up the gate; even locked the box. But Woytek still had a lot of people over that she didn't trust."

I raked my fingers through my hair as if they could untangle my guilt. "Yeah, she told me a bit about that, but I didn't realize. . . ."

"Listen, don't blame yourself." Jim squeezed my shoulder. "There's nothing you could have done. Roman wanted Woytek to stay there and Sharon never crossed his wishes." He knocked back the rest of his drink. "Man, she didn't deserve that schmuck. He treated her like a dog."

Wende blotted her eyes. "Do you know the worst of it, P.J.? I keep thinking if she'd married Jay, she'd still be alive."

"I guess that depends on whether you believe in fate or not," I replied.

I'D MET STEVE McQueen a handful of times. He was an intense character whom I kept a close eye on while he dated Sharon. He'd always been reclusive; following the murders, he went into complete seclusion at his estate, christened the Castle. Massive walls created a fortress around the two-story villa set an acre back from Oakmont Drive. McQueen wouldn't see me if I called or rang the gate. There was only one chance of getting through.

Two blocks from his house, I waited for McQueen's maid to arrive for work. At the base of Oakmont, a pickup truck dropped off three women. The smallest of the group walked to McQueen's driveway and used her key to activate the gate.

I edged the car forward until the thick, horseshoed gates accommodated my entrance. Just beyond, I passed beneath a broad, Moorish portal that housed security cameras. McQueen was out of the house before my car made it to the motor court. Shirtless and barefoot, the actor leaned onto the car door, trapping me inside. "Colonel, what brings you to this neck of the woods?" he stonily asked.

Clearly unwelcome, I got to the point. "I need to know where Jay scored his drugs."

His jaw clenched. "I don't have a clue."

"Come on, Steve, it's no secret. Give me a name."

"I never asked him."

"What did you find when you cleaned out Jay's house?"

"I didn't," he said tightly.

"Bullshit."

McQueen's eyes turned to steel. "I only made the suggestion." Then he indicated that other people did the job. As I attempted another question he leaned closer and said, "Colonel, you're a smart man, but you're barking up the wrong tree. You ought to look closer to home and check the dirt on Polanski's hands, because I'm not the one secretly checking my friend's cars for bloodstains. I know I don't have friends that could murder a pregnant woman. Apparently, your son-in-law isn't as confident."

He started away, terminating the conversation. "Damn shame you didn't make it to Sharon's funeral," I verbally chased him. "She obviously thought a lot more of you than you think of her."

He paused as if to turn, but then continued on his path.

I FOLLOWED UP with each of Sharon's friends. Over the phone, the innocence of Warren Beatty's voice countered his reputation in every way.

"Colonel Tate, Warren Beatty here. They said you called?"

"I did. How are you, son?"

"Not so good. I miss them."

There was a brief silence followed by a soliloquy in which his thoughts continuously flowed without pause. It was tough to keep up with what he said. But what did stick with me and always would was his last thought. "Hey, do you remember that year you had us all over for Thanksgiving? That night, I think you and Doris went to bed early or something. Anyway, we're in the backyard watching for shooting stars. It was a crystal-clear night. Nobody said much; we were just . . . together. Someone may have lit a joint, I don't remember. But I do remember we passed that night like so many others— blamelessly. Not like what they're writing in the papers. Jay was a good man; the best friend a person could have."

The subtext of his comment was clear to me: Drugs or no drugs, Jay didn't do anything to bring on the murders. And if someone cleaned Jay's house of drugs, it was a harmless act to protect his family.

The short time I'd been on the phone, I fiddled with a bottle of Sebring brand shampoo. Beneath the bottle lay Guy's disturbing report on Jay and Sebring International.

In spite of the growing business and his penchant for extravagance, he was more than a quarter of a million dollars in debt. In the last year, he'd borrowed money from everyone, $5,000 from Gibbie, $2,700 from Sharon, even $6,000 from his dentist. Was there a loan we hadn't uncovered?

The police found two vials of cocaine in his car, and tiny packets of the drug in his briefcase. When I finally met up with another of Jay's friends, Jim Coburn, he confirmed that there was coke stashed all over the house when it was cleaned. Jay was like a son to me; how could I have missed all the signs of the trouble brewing in his life?

My mind drifted to a better year, a different house, and the day that Jay proposed to Sharon. He'd done it by the book and arrived early the day before to ask my permission. He actually used the ridiculous words, "May I have your blessing to take your daughter's hand in marriage?"

I stifled a laugh, popped open a couple of beers, and told him, "Sure. What the hell."

The next day, he and Sharon came for dinner. In the middle of grilling steaks on the barbecue, Jay got down on one knee and popped the question to Sharon. As I watched the romantic spectacle, the urge to laugh was there again, but this time it was stifled by the thought that we'd gotten damn lucky to have Jay for a son-in-law. He was a good man and good to Sharon. What more could a father ask for?

My hand still rested on the phone when its ring brought me back to the present. "Tate here," I answered.

"Candy Bergen may know something about the murders." The line went dead.

Unlike some of the anonymous calls, I took this one seriously. Candice Bergen and Terry Melcher had lived at the Cielo house eight months before the murders.

Bergen had had many weeks to contemplate escaping Sharon's fate. By the time she agreed to a phone conversation, she'd calmed to terrified. Her call pulled us into the eddy of the gossip pool. She related a thirdhand rumor, a variation of Jim Mitchum's story on Billy Doyle. In her version, Doyle flew to Jamaica to buy a large amount of dope for Abigail and Woytek. When the money didn't arrive from

Los Angeles, Doyle returned in a rage over the blown deal. Packing a gun, he went to Cielo, threatening to kill Woytek. Jay helped Woytek overtake Doyle and together, they took him to Cass Elliot's house, tied him to a tree, then raped him in front of an audience of twelve. Doyle vowed revenge.

It was time to track down Doyle.

"P.J., don't waste your time," Helder said. "Doyle passed a poly-graph."

"I want to talk with him myself."

"You know I can't give you that information."

I started to argue, but Helder cut me off. "Listen, have you ever been fishing in the Toronto area? I've even thought of buying some property up there. You should check it out. My other line's blinking. Catch you later."

Preoccupied with anger over Helder's withholding information, I almost missed the tip.

FROM A PARKING lot across the street, Guy and I watched Doyle lock the door to his father's real estate office in Toronto, Canada.

"Is that him?" I asked.

Guy compared a picture of Doyle. "Yep."

"Smaller than I figured."

"Yep. They always are," Guy said.

As Doyle pulled out his car keys, Guy stepped forward. "Hi, Billy. I'm special agent Morreau and this is Colonel Tate." He dropped his sunglasses down his nose, squinting at Doyle. "You sure don't look like a freaked-out candy man."

The keys jangled in Doyle's hand. "That's because I'm not," he said, missing the car lock twice.

Guy reached for the keys. "Here, let me help you with those, you're gonna scratch the paint."

"I already told the police everything I know," Doyle whined.

"You're not going to cry are you, Billy? P.J., check his pants and make sure he didn't wet them."

Doyle leaned against the car, his arms wrapped across his chest. "What do you want?"

My role was the good guy. "We just want to talk."

"I've seen pictures of you. You're Sharon's father, right?"

"That's right. Why don't you let us buy you a beer?"

"I don't drink."

"Good, the tab will be cheaper," Guy said, firmly leading him by the arm.

We sat in the last booth of a gloomy pub. Doyle's fidgety moves squeaked on the red plastic bench while his fingers rolled a paper place mat. "Sharon was a lovely girl. She never took drugs or drank like they're saying."

"Well, there's your first lie, boy," I admonished. "I've come a long way to talk to you, and it sure as hell wasn't to listen to a bunch of fabricated crap."

Doyle's voice shifted to a clipping staccato. "I'm not lying. I never saw Sharon high. Ever."

"How well did you know my daughter?"

"Not well at all. I went to a party at her house in January. I saw her at some other parties. Once at Jay's, once at John Phillips's, and then I didn't see her again until July."

"I hear you crashed that party at Cielo and then threatened some folks," Guy said.

"I did no such thing. I was an invited guest. Woytek invited me, and Roman invited Cass. We were invited as a couple."

"What about the threats?" I asked.

"You've got the wrong man; that was Pic Dawson."

"If you're such a good guy, why's everyone pointing the finger at you?" Guy asked.

"This mess was started by John Phillips. As I'm sure you know, I was living with Cass Elliot until a few weeks before the murders. I love Cass; we have a lot in common. After the murders, John convinced her that I killed all those people. She believed it and told the police a bunch of false things about me. She has since apologized."

"Why did Phillips have it out for you? Did you ball his wife?" Guy quizzed him.

"No, but Roman did. Has anyone considered that John did this for revenge?"

"Let's concentrate on you for the moment," I told him.

"Michelle Phillips is a cobra. I couldn't be less interested in her. As for John, he's a violent fellow who liked to settle all of his arguments with Cass by hitting her. I stopped that, and he's never liked me since."

Doyle seemed too composed to Guy. He decided to shake him up and see what spilled out. "Tell us about the time Woytek tied you to a tree and balled you in the ass."

"I, uh, I—he didn't, Woytek invited me up to the house to do some mescaline. I only remember a bit of the party." Doyle stopped, rolled his eyes, defeated. "I can't give you the facts. I was unconscious. But I wasn't sore there the next day, and I was sore everywhere else."

"Tell us what you do remember," I said.

"Woytek and I had been drinking a lot of Champagne and we got into a disagreement. He was a Communist and bountifully anti-Nazi. I made the mistake of saying that Asiatic communism was the worst political system in the world. Woytek became hysterical, saying the Nazis were the worst. After that, I remember Woytek said he'd given me an overdose; not maliciously, just to kind of get back at me as a joke. I was really high and needed something to bring me down. He brought me some pills that he said were Sharon's—very mild sedatives. After I took about eight of them, he told me they were something else and started laughing. I went crazy, throwing things and such. They called my friend Charles Tacot. He took me to Cass's house and chained me to a tree for eight hours."

"Are you still dealing?" Guy asked.

"I never did. It's true, I told people that I had lots of coke—pounds of it—but it was all hype. All these famous people were having parties, and there'd be thirty naked girls—mostly groupies—who wanted to get laid, but their price was a nose full of coke and I wanted to fuck them all."

"Was this crap going on up at Cielo?" I asked.

"Only when Gibbie and Sharon were out of town, and then only for the five-week period that Woytek had run of the house."

"Who were you getting the coke from?" Guy asked.

"No one in particular, wherever it could be found."

"What about this big shipment of MDA everyone's buzzing about? Was Woytek setting up shop in L.A.?" Guy persisted.

"Woytek led a truly social life. Nevertheless, the rumor of Woytek being a drug dealer is ridiculous. He barely spoke English. Anyone who represents Woytek as being a bad fellow has something to gain or hide. He was a good man, a good friend, and he was never strung out the way they're writing about him. His other flaw was that he wouldn't marry Gibbie. He loved her, but hated her wealth."

"What about Gibbie?" I asked.

"She was madly in love with Woytek, but she would have left him if he didn't marry her."

"I meant, was she using?" I prodded.

"*No way*. This was a girl who thought it was incredibly mischievous to take a toke off someone else's joint. Perhaps she was different around others, but she didn't use drugs in my presence. It's a bunch of nonsense what they're writing in the newspapers about those girls. I didn't know Sharon all that well, but I knew Gibbie. They were lovely girls, Mr. Tate. Don't let anyone tell you differently."

Guy gave me a knowing glance. Another dead end.

SUBSEQUENT TO EVERYTHING I'd heard about Cass Elliot, I expected to find a brothel instead of a two-story Cape Cod tucked into layers of tropical greenery. A fountain splashed additional tranquillity; my knock on her door would have sounded intrusive even if I'd known the singer. I reached to knock again. "I heard you the first time," said a voice from behind the door. "Who are you?"

"I'm Sharon Tate's father. My name is Paul."

A security chain kept the door from opening further than an inch. "They said you might come here. I've already told the police everything I know."

"Sometimes the police don't ask the right questions. If I could have just five minutes of your time."

Elliot stepped back. I heard a pistol being disarmed; by the sound of the slide action, I guessed it to be a Walther PPK.

"I'm sorry about Sharon," Elliot said, opening the door.

We sat in rocking chairs on the front porch. For such a large woman she had a tiny and frail demeanor. "Sharon murdered right in her own living room . . . I'm scared—beyond scared—I jump if my neighbor coughs."

"How did you meet Sharon?" I eased her into the subject.

"Steve Brandt—I guess you know he tried to commit suicide?"

I nodded. Brandt did publicity work for Sharon and other entertainers. His attempted suicide raised a flag of suspicion; but his stories were so fantastic that the LAPD dismissed him as a possible suspect or witness.

"He planned all of Michelle and John's parties," Elliot continued. "He must have invited Sharon and Roman to one of them. After that, Roman was up at their house a lot more than Sharon. He didn't like me so much after I found out about him and Michelle. Strange guy. It didn't matter if Sharon knew, but when it came to John, that was another story."

"Did you tell John?"

"No. By that point, we had enough problems in the group. John would have turned it around and made it my fault. He did find out though; Michelle told him during an argument."

"Billy Doyle said John was pretty violent. Do you think he's violent enough to murder Sharon for revenge on Roman?"

"I just don't know anymore. Everyone suspects everyone. I've watched friends turn on friends over this. It's crazy."

"What made you think Doyle and Dawson were the killers?"

"I didn't. It was Brandt and John. They said the 'pig' inscription on Sharon's door was actually Pic [Dawson]. John blamed me for bringing killers into our group, blamed me for Sharon's death, and said the police were going to arrest me as an accomplice. I was mortified. They completely convinced me that my daughter and I would be murdered next. At that point, I would have said anything to get them behind bars."

"So there's no one you suspect from the Canadian group?"

"No. Contrary to the gossip, I didn't invite a bunch of drug-dealing strangers to come live with me. I'd known these guys a long time. Granted, neither of them are angels, but I don't believe they're killers. I've probably ruined their lives by telling the police that they were."

"In all the parties that you went to, is there anyone that didn't seem to belong, any event out of the ordinary?"

"What *was* strange? Who didn't fit in would be what you consider the norm. It was a constant party at John and Michelle's. A hundred people could pass through when we recorded there and I wouldn't know seventy-five of them."

Cass's head drooped. "In the beginning we were dreaming about making miracles. The music, the money, the drugs, even the weird behavior—it all seemed right. I don't even know how it all went so wrong, but I guess it's over now, isn't it?"

AT THE SAME time I interviewed Cass Elliot, Frankie made an unannounced visit to John and Michelle Phillips. Later, we met at the top of Fryman Canyon to compare notes.

"How'd it go?" I asked when Frankie pulled up.

He let out a long whistle. "We're talking out there. My greeting was a half-dozen security guys accosting me while Phillips hid behind them like a bullied kid in the park."

"What's he so afraid of?"

"Well, that was kind of interesting. He swears that Doyle and Dawson are the killers and they're coming after him for making them to the cops. He claims that he got to know Folger and Frykowski through Polanski. Spent a lot of time at the Cielo house while Sharon was in Italy. Says that when Folger was present the place was pretty mild, but when she wasn't, there'd be twenty or so naked partiers doing mass quantities of drugs—drugs supplied by Doyle and Dawson. They might be married, but she's carrying on with half of Hollywood, and he's doing the other half and proud of it—even offered to show me his personal sex videos.

"The missus followed me out to the car. When I asked if she had something on her mind, she seemed very concerned that the police considered her husband a suspect and wanted to know why. So I asked her straight up if there was any reason Phillips would have any animosity toward anyone at Sharon's house, and that's when she admitted Phillips found out about the affair she had had with Polanski in London—feels really bad about it now with Sharon dead and all."

"Anything else?" I asked.

"Nah. It was tough to keep a straight face while he denied using drugs himself because the place reeked of grass. First, it's Dawson and Doyle, then he says it's the Mafia and he's next on their murder list— hit men have been following him. Then he babbled about drugs, psychics, and some guy out in the desert that steals your soul. I couldn't keep it all straight. If he knows anything, it's lost in whatever's left of his brain."

RATHER THAN WASTE hours sleeplessly tossing and turning the night away, I ran surveillance at the Cielo house in case the killer returned to relish his crime scene. Above Cielo Drive, another cul-de-sac provided a vantage point of the entire neighborhood. Through field glasses, I'd watched dozens of cars trek up to the gate. From Mercedes to Datsuns, the curious came from every lifestyle.

Midnight struck while I stretched my legs under the moonless sky. Coyote yelps traveled through the wind's careless scattering of sounds, and then the faint roar of Harley-Davidson motorcycles reverberated from the canyon base. Shadowing the echo, a pair of headlights rolled side by side up Benedict Canyon Road. I tracked the bikes to Cielo.

Different from the other curiosity seekers, they operated the Harleys with the confidence of knowing their destination. They parked at Sharon's gate; two dismounted from each chopper. With their long hair, it was tough to tell the men from the women.

They climbed the chain links, vying for a better view, and seemed to be contemplating going over when two guard dogs bounded around the corner of the garage. The aggressive dogs should have sent the foursome scrambling, yet they stayed, taunting the animals into a frenzy. It wasn't until the neighbor's light illuminated the street that they bolted.

There is only one way in or out of Benedict Canyon, south into Beverly Hills or north toward the valley. I gave them a half-mile lead before following their trail north.

There is never a quiet time on the Los Angeles freeway system, including the one o'clock hour that I mounted the eastbound ramp of the 101 freeway in Sherman Oaks. I closed the gap between the Harleys as we headed deeper into the valley, toward Ventura. Thirty miles into the trip, they exited onto Topanga Canyon, riding toward desert terrain.

Past the suburban area, the road was desolate, forcing me to cut my headlights. Ahead, the motorcycle lights sliced through the blackness, assaulting the serene landscape as they made a sharp left turn, then another, before dipping into a basin.

The first violet glints showed in the sky before I chanced the turn down the isolated Santa Susana Pass. I matched the speed of the bikers, and then counted their seventy-five-second trip. At the next driveway, I stopped near a splintered wooden sign with a sun-faded inscription: SPAHN'S MOVIE RANCH. Silhouetted beyond the placard, was a mock western town. Cars, Jeeps, choppers, and psychedelic signs infiltrated the Wild West boardwalk.

For a better look at the layout, I backtracked to Topanga, then up Devil's Canyon Highway to the first plateau above Santa Susana Pass.

To the right of the western town stood a rundown house opposite corralled horses that looked as uncared for as the auto shells, rusty appliances, toppled coaches, and trash that filled the remaining areas. Behind the main boardwalk were trailers. Beyond the trailers, shacks and tents were scattered in the rocky foothills.

The rising sun turned the evening's frostiness into a furnace, recharging to hit its noon zenith. A mélange of people—from toddlers to the elderly—began emerging from the buildings and tents. Apparently, communication wasn't part of the morning ritual, as they dispersed around the ranch without so much as a nod to one another. Some of them looked the part of a ranch hand, while others seemed wayward in hippie garb. Contrasting to a greater degree were the ones who didn't bother with clothing. There was, however, one commonality among the residents. The toilet was an ignored contraption.

What in the hell is this place?

Jake and Guy were making breakfast when I got home. "Just make yourself at home, gentlemen," I said.

"What," Jake answered, "we're gonna wait for your booty to come home? Where the hell you been?"

"Following some bikers that showed up at Cielo last night."

"Anything?" Guy asked.

"I don't know. There were no tags on the bikes, so I trailed them to an area out in Chatsworth—Spahn's Movie Ranch—strange place."

"Call me crazy," Jake said, "but wasn't there something in the newspaper back in August about that place being raided?"

I shrugged, not remembering.

"Let's put a call into the sheriffs out there and see if they got anything," Jake suggested.

"Not until I've showered," I said. "I'm filthy just from looking at the place."

"Princess, you go ahead. Let the real men take care of the work," Jake winked at Guy.

On my way out, I saluted them with the finger.

After my shower, I fell soundly asleep. I didn't know Jake and Guy had left until the phone rang. "Hey, sleeping beauty, you awake?" Jake asked.

"Getting there, what's up?"

"I'm gathering the troops to your place. I've got something hot."

THE NIGHT FOLLOWING Sharon's murder, a middle-aged couple were slain in a strikingly similar manner. At a glance, Leno and Rosemary LaBianca had nothing in common with Sharon's crowd. They lived a quiet life in the conservative suburb of Los Feliz. The couple's sole connection to the entertainment industry was once living in Walt Disney's former home. Leno was president of Gateway supermarkets. Rosemary owned a dress shop in downtown Los Angeles. The La-Biancas' paths never crossed Sharon's, and yet they were intrinsically tied together by their murders.

The team sat around the kitchen table poring over the LaBiancas' detective's second progress report. Jake was in the middle of his discovery. "So I call up the sheriffs and get a deputy who tells me that on August sixteenth they raided this Spahn's Ranch to bust an auto-theft ring. The raid turns out to be a cop's wet dream; they uncover an arsenal of weapons, drugs, stolen credit cards, IDs, abused minors—you name it. After the arrest, they have to release the entire group because the warrant stank. Half of them—including the key guy, Charles Manson—skip town to another ranch out in Inyo County."

Jake turned the page. "Now, check this out, at the bottom of the LaBianca suspect list is Charles Manson. Why him? Because back in July the sheriffs had a similar murder. Their victim, Gary Hinman, also had bloody messages on the wall—including the word *pig*—and the same stabbing overkill.

"The sheriffs bust Bobby Beausoleil for Hinman's murder; he's driving the vic's car with a bloody knife in the trunk, fingerprint match from the scene, the whole shebang. Now, Bobby Beausoleil's a cat heavily tied to Manson and his group.

"So, I call Sergeant Guenther, the lead on Hinman, and he says he talked to Helder's boy, Jess Buckles, the day of Sharon's autopsy; tried to convince him he's got a link to one of his killers in jail. Buckles, the cocky prick, says, 'We're working a major narcotics ring. Our case has nothing to do with a bunch of hippies.'

"The next day, after the LaBiancas are killed, Guenther goes to those detectives with the same spiel. They disregard Guenther as a no-brain hick."

"Jake, are you going to take a breath anytime soon?" I interrupted.

"What?" Jake looked at me innocently.

"You lost me about five minutes ago. Explain how Manson ends up on the LaBianca report and the tie-in to Sharon."

"Just hear me out. Hinman had two cars. In October, the sheriffs raid the Inyo County ranch, find Hinman's other car, and arrest twenty-four freaks, including Manson. In jail, the chicks from the group start spilling their guts about murder. One in particular, Kathryn Lutesinger, is scared; says that they're going to kill her because she knows about the Hinman murder. So she farts on all of them, giving the names of three who killed Hinman: Beausoleil, Sadie Mae Glutz, and Mary Brunner—Manson was there, but he didn't participate.

"During Guenther's interrogation, Lutesinger tells him that one of the girls stabbed Hinman in the legs. Thing is, Hinman wasn't stabbed in the legs—but Woytek Frykowski was."

I raised my eyebrows in frustration. "The point, Jake?"

"Yeah, yeah, I'm getting to it. All along, this guy Guenther has a boner to tie all three cases together—Hinman, Sharon, and LaBianca—and starts wondering about the vic with the leg wounds. He calls Helder's group again, gets no return call. Then he calls Jimenez from LaBianca, who's a little more humbled by this point, so he throws Manson's name in the hat."

Jake turned the report to page fifteen. "Take a look at this. The LaBianca detectives start putting two and two together and run an MO for suspects who wear glasses and disable phone lines—that's Sharon's case, not theirs."

Jake tapped his heart. "My gut instinct's telling me these three cases are tied to the same perp. I ain't Sherlock Holmes, but I ain't stupid, either. If Manson and these other clowns did Hinman, then they did Sharon and the LaBiancas."

As an afterthought, Jake added, "Forgot to give you the frosting. The bikers you followed? Probably Straight Satans from Venice. Big-time bad boys. Guenther says one in particular hung out at Spahn's, goes by the name Donkey Dan."

BOB HELDER AND I shared a table at the Hamburger Hamlet. "Bob, the LaBianca detectives are trying to steal your thunder."

"What are you talking about?" Helder asked.

"Just what I said. One of your boys ought to lean across that long desk they all share and check it out. They're tying Sharon's case to the LaBianca couple without consulting your guys, and they're throwing away solid leads."

I slid the LaBianca report across the table. "I've seen it, P.J."

"Have you really, or did you scan it?"

"Scanned it," Helder admitted. "Not my case."

I turned the report to the MO run. "Wore glasses, or disabled the phone. Sound familiar?"

Helder took the report from me. "What else you got?"

"The last two names on their suspect list, Charles Manson and Kathryn Lutesinger, both tied to Robert Beausoleil, the suspected murderer of Gary Hinman."

"I know all about Beausoleil and Hinman. McGann checked it out. Beausoleil was in jail the night Sharon was killed."

"Yeah, but McGann didn't follow through on the lead. Manson, Lutesinger, and the rest of this group hanging out with Beausoleil were free as birds to kill Sharon. Up in the Inyo County jail, Lutesinger is begging to turn evidence before she's the next victim."

Helder pensively paged through the report.

"Bob, you're a damn fine detective, but the men under you need a wake-up call. I'm giving your guys a few days on this and then we're moving in."

We never got the chance. Within the week, confessions were rolling in faster than Helder could document them.

7

A FAMILY LIKE
NO OTHER FAMILY

When you have eliminated all which is impossible, then
whatever remains, however improbable, must be the truth.

—SHERLOCK HOLMES

"You could hide the Empire State Building out there and no one
could find it," commented a ranger on the complex geology of the
3.3 million acres of California's Death Valley, where temperatures rise
as high as 134 degrees and then drop to 30 degrees in the thick of the
night.

Naturists love the area for its vast and often colorful beauty. The
1840s' gold rushers who died crossing the treacherous range understood it only as its namesake. Charles Manson and his group found
the barren region an unyielding utopia of obscure places to disappear
in after the August 16 sheriff's raid on Spahn's Ranch. And had it not
been for the Inyo County authorities, the killers might have avoided
detection for years, possibly forever.

The Manson Family, as they called themselves, began appearing
on the desert authority reports as early as August 21, 1969, when a
Lone Pine deputy cited family member Charles Watson for loitering.

Game Warden Vern Burandt filed field reports from residents who complained about hippies killing the quails, defecating in public, and swimming nude in the creeks.

In the town of Shoshone, just south of Death Valley, Deputy Don Ward confronted a group he suspected of pushing hashish in the community. On Main Street, where Ward stopped to question them, Manson threatened, "You just made a big mistake, partner."

All those incidents aside, it was Manson's decision to burn the county's $35,000 earthmover that launched Inyo's search for the Family. On September 29, when park workers arrived at the sight of the still flaming machine, they noticed a set of Toyota 4×4 tire tracks. Desert authorities are well-bred trackers; the tire marks were fingerprints in the sand.

Six highway patrolmen, four deputies, three rangers, and airplane pilots pursued the suspects through Death Valley, from the northeastern shadows of the Nevada border to the southern San Bernardino county line. Over the span of five days, the trackers uncovered camouflaged dune buggies, cars, caches of auto parts, fuel, weapons, and food hidden by Manson's group.

Near the Inyo/San Bernardino border, Officer James Pursell and Ranger Dick Powell traveled into ruinous Goler Wash to check out two properties: the Myers and Barker ranches. Except for the occasional prospect miner who drifted into the area, both houses were vacant, filthy, and rarely inspected by the owners.

A quarter mile from the Barker acreage, the officers encountered seven seminude teenage girls, two men, and a red Toyota 4×4 with the suspected tire tread. The smaller of the two men took off into the brush. The taller one stayed and identified himself as Charlie Montgomery (true name Charles Watson).

"Who owns the 4×4?" Pursell asked.

"I have no idea, Officer," Watson smiled. "I'm from Olancha. I hitched a ride with these guys to get to the 395 freeway."

"Far cry from the 395," Pursell commented. "What are you doing here?"

"Needed a place to stay the night," Watson said.

Pursell looked at the women. "What about you ladies?"

One of them mumbled something incoherent that caused the other girls to laugh.

During Pursell's interaction with the group, Powell kept his distance, surveying the area for the smaller man who'd fled. From the corner of his eye, he caught some movement in the brush, and then the barrel of a shotgun. Outnumbered and out of radio range, he had to think quickly. "Come on, Jim, I'm hungry. These guys aren't causing anyone harm." When Pursell turned to argue, Powell shifted his eyes toward the brush.

Pursell caught on. "I guess you're right. Besides, I want to get down the wash before sunset. You folks take it easy."

Because of the rugged terrain, it took Powell and Pursell three hours to reach the town of Ballarat, where they made radio contact. By the time they returned with the support officers, the ranch areas were deserted, vehicles and all.

For a vastly uninhabited area, word of mouth spreads quickly through the Valley, and on October 8, the rangers received confirmation that the hippies had returned to the Goler Wash ranches. The prospectors who had spotted the group also gave the rangers a warning: Manson had armed patrol members scattered throughout the wash.

Acting quickly on the tip, four Inyo County law enforcement agencies organized a raid on the ranches.

There are numerous passages surrounding the Barker and Myers ranches, including Anvil, Sourdough, and Willow Springs, Mengel Pass, and Goler Wash. Between two and three in the morning on October 10, officers were dropped at each of the access points to hike the four miles into the ranch areas. On their trek, they arrested five armed Family members lying in wait.

The sun was well above the horizon by the time the officers descended on Barker Ranch. On the south ridge, above the main house, they spotted a camouflaged lookout post alongside two bunkhouses.

Teams deployed around each of the smaller dwellings and the

main house as another team drove a Jeep right up to the porch. Two women bolted from the front door and right into Pursell's grasp. A deputy who entered from the back brought out a third woman.

On the ridge above, the officers captured three girls in the lookout, and two more in the bunkhouses.

Over the half-mile radius between the Barker and Myers ranches, they apprehended ten women, two infants, and three men. At the time, authorities had no idea there were killers among them, including Leslie Van Houten, Susan Atkins, Patricia Krenwinkel, and Steven Grogan.

Atkins must have sensed that her freedom was about to end for a long time to come. She called over Pursell. "Would you unhook me for just a few minutes?"

He looked at her curiously. "Why?"

Atkins pointed toward Grogan. "I want to make love to him one more time. He's about the best piece I've ever had, and I may never see him again."

"I don't think so," Pursell told her.

"Fine," Atkins said, then squatted to urinate near his feet.

"Jesus Christ, what's the matter with you?" Pursell exclaimed at the sight.

Later that same evening, two young women approached Highway Patrol officers Anderson and Hailey, who gathered evidence from the stolen dune buggies recovered during the raid. Stephanie Schram and Kathryn Lutesinger, both Manson Family members, were frightened. "Can you help us?" Lutesinger timidly asked.

Anderson looked the two over. "What's the problem?"

"The people up at the ranch want to kill us," Lutesinger told him.

Hailey stepped in. "Why's that?"

"Because we wanted to leave," Lutesinger said.

"Well there's nothing to worry about now. They're all locked up over in Independence," Anderson said.

Lutesinger shook her head. "No, there's a lot more of them, and when they come back, they'll chop off our heads like they did Gary's."

Stephanie shot her a look.

"Who's Gary?" Hailey asked.

"It's nothing," Lutesinger quietly replied. "Can you just call our parents?"

Once in Independence, Officer Anderson did more than call the girls' parents. He researched murder victims named Gary. As it turned out, the Malibu sheriffs had issued a warrant for Lutesinger as a material witness to Gary Hinman's murder.

Based on Lutesinger's concern that Family members would return to the area, the deputies planned to raid the ranches again on October 12.

This time, the officers had the upper hand as they now knew the layout of the place. Pursell and Powell busted through the back door of the Myers Ranch house. In the kitchen they found three women and six men, including Charles Manson and Bruce Davis.

Although Inyo County is technically the second largest in California, their jail was tiny and now bursting with suspects. Working with the district attorney's office, the sheriffs processed the group, determining who they had enough evidence against to file charges for auto theft and arson and who could be let go. Unaware that they were housing a contingency of murderers, they released Bruce Davis, Patricia Krenwinkel, and Mary Brunner.

It could be said that Kathryn Lutesinger triggered the avalanche of confessions that came down around the Manson Family after she implicated Susan Atkins as an accomplice in Gary Hinman's murder. But there were others willing to disclose their knowledge; some of the informants lived to testify at the trial; others weren't so lucky.

When confronted by Sgt. Guenther, Susan Atkins, also known as Sadie Mae Glutz, surprisingly confessed to participating in Hinman's murder.

At the coffee shop in Independence, Atkins ate a sweet roll. In between bites, she recounted Hinman's three-day torturous execution. Lutesinger had it wrong; they didn't behead Hinman, just chopped off his ear.

Booked on suspicion of murder, Atkins was moved to Los Angeles and processed into the Sybil Brand Institute (SBI) for Women.

Rather than housing inmates in individual cells, most of the SBI facility had dormitory-style floor plans with more than two hundred women in each of the sleeping quarters. The open setting, including row upon row of bunk beds, made it easy for the women to have sexual relations after lights-out; a situation that Atkins eagerly took advantage of.

At the close of her first week's stay, Sexy Sadie was the talk of dorm 8000. The women joked, "She's the fastest tongue in the West and responsible for the removal of more underwear than last call for laundry."

By the time Atkins made her way to Ronnie Howard's bed, Sexy Sadie's promiscuous celebrity was fading, so she developed a new act to premier that night.

Different from the other women, Ronnie didn't push Atkins away after their sexual encounter; she wanted to talk, and Atkins obliged. Cross-legged at the foot of the bed, Atkins asked, "What are you in for?"

"Violating my parole. You?"

Atkins smiled. "First-degree murder."

"Did you do it?"

"Sure, but it's not the only one, there's more." Atkins's voice rose excitedly. "Are you ready for me to blow your mind?"

"Shush, there are ears all over this place. My ex-husband was busted for a robbery because he couldn't keep his mouth shut in jail."

"I'm not worried." Atkins rubbed Ronnie's leg. "I know that I can tell you things. Besides, if I ever got caught, I'm really good at playing the crazy little girl, plus I've got an alibi. None of that matters, though. The police are so far off track on the case, they don't know anything."

"What case?" Ronnie asked.

"The one in Benedict Canyon."

"Sharon Tate?"

"That's the one. You know who did it, don't you?" The murderess moved just inches from Ronnie's face. "Well, you're looking at her."

Ronnie lifted her head. "Why?"

"We wanted to do a crime that would shock the world."

"But why did you pick Sharon Tate's house?"

"Because it's isolated."

"Did you do it by yourself?"

"Oh no. Katie, Charles, and Linda helped."

"How did it go down?"

"We walked up to the gate, and Charles cut the telephone wires so that—"

"Wasn't he worried that he'd cut the electricity by mistake and be electrocuted?" Ronnie interrupted.

"No, he knew just what to do because we had visited the guy who lived there before. Anyway, just as we got over the gate, headlights shined toward us and before I knew it, I heard, pop-pop-pop-pop and—"

Ronnie stopped her again. "Is that the boy that was found in the car?"

"Yes, he was the first to die. Inside, we found Frykowski sleeping on the couch, and Abigail was in the bedroom reading a book."

"Wait, I thought you picked the house at random? How did you know their names?"

"I didn't until the next morning on the news, and boy, I knew them then." Atkins laughed. "Sharon and Jay were in the other bedroom, and when I went back to get them, they looked up, and were they surprised!"

"Did they fight back?"

"Are you kidding? They knew we meant business." Atkins sat up, her voice enthusiastic. "Frykowski broke free and ran for the door. I stabbed him three or four times; he was full of blood and bleeding all over the place. He ran to the front door and out onto the lawn— and would you believe that he was hollering, 'Help, help, somebody please help me!' and nobody came." Atkins giggled. "We finished him off in the yard.

"Sharon was the last to die. I had her arms pinned behind her back and she was begging and crying, 'Please don't kill me. I don't want to die. I want to live and have my baby.' I stared her right in the eye and said, 'Look, bitch, I don't care about you. I don't care if you're going to have a baby. You had better be ready because you're going to die, and I don't feel anything about it.' Then in a few minutes, I stabbed her. It felt so good, and when she screamed, it sent a rush through me, and I stabbed her again. It's like a sexual release, especially when you see the blood spurting out. It's better than a climax." Atkins reached for Ronnie's hand. "Have you ever tasted blood?"

Too afraid to utter a word, Ronnie shook her head.

"After Sharon was dead, I thought about cutting out the baby. I had blood all over my hands, and it was so warm and sticky, and nice, so I tasted it. Wow, what a trip, to taste death, and yet give life!" Atkins's head fell back to the pillow. "You know the couple that was killed the next night?"

"The ones in Los Feliz? Was that you and your same friends?"

"What do you think? And there's more; there are at least eleven bodies that they'll never find."

A career criminal, Ronnie had never snitched on anyone, but this was different. This was insanity, and she was going to break the code of silence.

In the meantime, the murders continued.

On November 5, the Venice police responded to an emergency call on Clubhouse Drive. John Haught was dead from a gunshot wound to the temple. Bruce Davis and other Manson Family members claimed Haught shot himself playing Russian roulette. There was an oddity noted in the police report that didn't tally with their story: "The eight-shot, .22-caliber Iver Johnson revolver held seven live rounds and one spent shell."

On November 16, the body of a young girl was found in the Santa Monica Mountains where Mulholland and Bowmont Drives intersect. She'd been stabbed 157 times. Lacking identification, the body was taken to the morgue and tagged Jane Doe 59. Spahn's Ranch manager Ruby Pearl later identified the woman as "Sherry," a girl who hung around the ranch and the Family.

November 17 was a big day for the LAPD. They were literally handed the solution to the murders from numerous sources.

Ronnie Howard had a court appearance, and a chance to make a phone call to the LAPD. Sgt. Larry Brown, received the call and then phoned Helder. "Bob, I got a paperhanger down at SBI says she's solved the Tate murders for you," Brown laughed. "Want me to check it out?"

"What the hell, it can't hurt. Call me here if anything turns up."

That same day, based on a tip from Sgt. Guenther, the LaBianca detectives interrogated Leslie Van Houten at the Inyo County jail. Van Houten admitted that members of her group, specifically, Tex, Katie, Sadie, and Linda, might have been involved in the Cielo Drive murders. As for the LaBianca couple, she claimed ignorance.

And while the Van Houten interview was in progress, a Venice Beach detective called Bob Helder. They had a biker in custody. A Straight Satan, Alan Springer, wanted to trade information on "Tate."

Helder sent Mike McGann to interview Springer. Because a majority of what Springer relayed came secondhand from Satan treasurer Daniel DeCarlo, McGann had little faith in Springer's information, until the biker let go of a bombshell. "Hey, Charlie said that they wrote something on the refrigerator in blood."

McGann was familiar enough with the LaBianca case to know that the killers had written "Helter Skelter" on the refrigerator at that scene. A clue harbored from the public.

Suddenly interested, McGann asked, "Where's DeCarlo?"

"He's hiding, scared these freaks are going to kill him," Springer said. "You get me outta here and I'll bring him in for you. He'll back up everything I'm saying, man."

Springer was released on his own recognizance, but his freedom was contingent on bringing in Dan DeCarlo.

That same evening, Springer arrived with DeCarlo, and in a small interview room, Donkey Dan pinned the Cielo Drive, LaBianca, and Hinman murders to Charles Manson and the Family. DeCarlo was the "weapons man" at the ranch. He cleaned the guns and sharpened the knives, so it was easy for him to link the evidential murder weapons from each of the crime scenes to the Family.

During the course of the DeCarlo interview, the death toll grew. He told the detectives that Manson ordered the murder of ranch hand Donald O'Shea because he'd been snitching to George Spahn about the Family's thefts and drug use.

Helder had just finished the DeCarlo interrogation when Larry Brown called from SBI. "Bob, you'd better get your ass down here. This gal knows what she's talking about." Details of Ronnie Howard's story were ringing true.

The next morning, nine detectives and two lieutenants, including Bob Helder, gathered in the office of Roger Murdock, the chief of detectives. Murdock gave a bittersweet smile while the detectives ran down the specifics of the case mounting against the Manson Family.

There was one glitch in the "case solved" celebration: Except for Charles Manson and Susan Atkins, the authorities only had aliases or first names of their suspected killers: Charles/Tex, Katie, and Linda.

"Bob, how do you see proceeding?" Murdock asked.

"We've got corpses piling up all over the place. I say we get the DA's office involved and get the ball rolling."

VINCENT T. BUGLIOSI joined the District Attorney's Office in 1964, just after graduating from UCLA. A deputy district attorney for five years, he was viewed as the quintessential prosecutor due to his hands-on investigative approach as well as his drive to uncover the facts.

True to form, Bugliosi was at Spahn's Ranch within eighteen hours of his assignment to the Manson Family case; collecting evidence, along with Lt. Helder, Sgt. Guenther, and their guide, Dan DeCarlo.

During a conversation with one of the ranch hands, Helder learned the identity of the suspected killer Linda. Her last name was Kasabian. Beyond that, the search was a bust because of gusting winds. It wasn't the day's only disappointment for the prosecutor.

Police Chief Ed Davis waited with Asst. DA Busch for Bugliosi's return. "How did it go out there?" Busch asked Vince.

"Miserable. The winds—"

"We talked to Susan Atkins's attorney this morning," Davis interrupted. "He wants to deal; immunity in exchange for her testimony at the grand jury hearing. I think we should bite and wrap this whole thing up."

"You're kidding, right? There isn't a case to take to the grand jury! I don't even know for sure who the killers are—or their real names, for that matter; this is way too premature."

"All the more reason to make the deal with Atkins; she'll supply the missing pieces. The public pressure is too big not to move on this." Davis smoothed back his already slick hair. "The press is on to this Manson Family, and they're only willing to hold their stories until the first of December."

"If what Atkins told Ronnie Howard is true, she personally stabbed Sharon Tate, Woytek Frykowski, Gary Hinman, and who knows how many others. This is not someone you offer immunity to!" Vince scornfully told the chief.

Busch intervened. "Vince, what if they kill again between now and December first? We don't have a choice."

In spite of the efforts of the combined law enforcement agencies, two more lives were taken on November 21. At 11:30 that evening, the bodies of James Sharp and Doreen Gaul were found in an alley near downtown Los Angeles. Both bodies were bludgeoned; both had at least fifty stab wounds. Coincidentally, Family member Bruce Davis had previously dated Doreen Gaul.

Together, the detectives and Vince moved at lightning speed to gather evidence that would tie the suspects to the August 9 and 10 murders. Within the two-week deadline, more than a hundred people were interviewed; one led to their first piece of physical evidence. It turned out that Manson was an aspiring musician who'd befriended Beach Boy drummer Dennis Wilson and subsequently Terry Melcher, a record producer and former Cielo tenant.

After an unsuccessful interview with Melcher, Sgt. Patchett from the LaBianca team interviewed Melcher's business associate Gregg Jakobson.

Impressed by Manson's philosophy and music, Jakobson told Patchett, he decided to introduce Manson to Melcher. "Charlie's views about life were different from anything I'd heard before," Jakobson said.

"Yeah, well, considering what I saw at the LaBiancas' house, you ought to be grateful for that."

"Man, you don't understand what I'm—"

"Mr. Jakobson, right now, I'm trying to keep anyone else from learning about Manson's philosophy on life. So how about you just answer the questions I ask. Ever heard of Charlie Montgomery?"

"That's Tex, I know him pretty well. His real name is Charles Watson. I think he's back in Texas."

Patchett ran the name through their database and found that Charles Denton Watson had been arrested in April of 1969 for possession of drugs—he'd been booked and fingerprinted. Watson's right ring finger matched a print lifted from the Cielo front door.

Twenty-one hours before Chief Davis's planned announcement, the state had their sole piece of physical evidence against the suspects.

In the rural community of Denton, Texas, just a few miles from the suspected killer's hometown, Sheriff Tom Montgomery arrested his cousin Charles Watson for murder.

The Malibu sheriffs provided the missing link to the suspect known only as "Katie." During the August 16 raid at Spahn's Ranch, they had arrested a young woman going by that name. Her father, Joseph, arranged for her release. At the time, he'd signed her out as Patricia Krenwinkel.

While the media set up cameras and microphones in the press auditorium, authorities pursued Patricia Krenwinkel. Her father gave detectives her address in Mobile, Alabama.

At the same time the LAPD phoned Mobile with a warrant for her arrest, Krenwinkel called his daughter to ask why the police wanted to locate her.

Ten minutes before Chief Davis's live broadcast, Alabama authorities apprehended Krenwinkel a block from her aunt's home.

"Ed, you can't go through with this," Bugliosi announced at 1:30 in Davis's office. "I don't have enough evidence on Manson to get an indictment. If it wasn't for the arson and auto theft charges, he'd be on the loose. I've barely scraped together enough evidence to get the warrants on Krenwinkel and Kasabian, but they won't hold."

"There are over two hundred reporters from all over the world waiting out there. I don't have a choice," Davis argued. "You can join me or not, I'm going out there."

Chief Davis didn't extend the same invitation to the Inyo County authorities who had worked tirelessly to capture the killers and break the case.

Seated behind a spray of microphones, Davis squinted into the bright camera lights. "After 8,750 hours of tenacious investigation by the LAPD, warrants have been issued for the arrest of three individuals in connection with the murders of Sharon Tate, Abigail Folger, Woytek Frykowski, Steve Parent, and Thomas John Sebring. These persons were also involved in the murder deaths of Rosemary La-Bianca and Leno LaBianca. The development of information from these two separate investigations led detectives to the conclusion that the crimes in both cases were indeed committed by the same group of people.

"The persons for whom warrants have been issued are: Charles D. Watson, twenty-four years of age, now in custody in McKinney, Texas; Patricia Krenwinkel, age unknown, now in custody in Mobile, Alabama; and Linda Kasabian, not yet in custody."

Davis didn't mention there were other Family members who were still at large and hunting their victims.

At about the same time the press conference took place, Joel Pugh, husband of Family member Sandra Good Pugh, was found dead in a Talgarth Hotel room near the Kensington (Olympia) station in London, England. His throat and wrists had been slashed. As with so many other murders committed by Manson Family members, the killer left a bloody inscription.

"IN THE EVENT that she testifies truthfully at the grand jury, the prosecution will not seek the death penalty against her in any of the three cases. . . ."

It was the deal of a lifetime for Susan Atkins.

Twenty days before Christmas 1969 reporters trampled over one another vying for a glimpse, a photograph, or a statement from Susan Atkins.

Wearing a rose-colored velveteen dress, her dark, shoulder-length hair flipped at the end, Atkins smiled demurely one moment then cackled the next, her character changing with the flash of the cameras as she posed for the newsmen. "Would you like to have a news conference with us, Susan?" one reporter yelled over the others.

"Sure!" Atkins said.

"What would you tell us?"

"Exactly what you want to hear!" she said, with the glee of a child.

Picked by lot, the twenty-one members of the grand jury formed a diverse group; nevertheless, they each viewed Atkins on the witness stand with the identical perspective. Her attitude was appallingly callous. Perfectly clear, her voice was sometimes flat, other times excited, but never remorseful as she verbally sketched two nights of murder with a complete disregard for human life.

Atkins's nearly black eyes widened at Vince Bugliosi's first question. "Are you a little nervous?"

Atkins crossed her legs, elegantly rested her hands on her knee, and in a voice of saturated composure, replied, "Scared to death."

Once Vince had established that Atkins met Charles Manson in 1967 while living in San Francisco, and he'd provided some details about the evolution of their group, he asked, "During this period, were all of you girls Charlie's girls, so to speak?"

"We were called Charlie's girls, but Charlie often told us, 'You people do not belong to me, you belong to yourself.' "

"What was it about Charlie that caused you girls to be in love with him?"

"Charlie is the only man that I have met on the face of this earth that is a complete man. He has more love to give to the world than anybody I have ever met."

"Susan, do you think Charlie is an evil person?"

"In your standards of evil, I would say yes. Looking at him through my eyes, he is as good as he is evil. You could not judge the man."

"Did you live at the Spahn's Ranch in Chatsworth with Charlie Manson and the other girls?"

"Yes."

"What type of life did you lead on the ranch?"

"It was beautiful, very, very peaceful. We took care of the ranch and each other. We all made love with each other, got over our inhibitions and inadequate feelings, and became very uninhibited."

"Did you call your group by a name, Susan?"

"Among ourselves we were called the Family—a family like no other family. We loved the whole world completely."

"Susan, on the date of August 8, did Charlie Manson instruct you and some other members in the Family to do anything?"

"I never recall getting any actual instructions from Charlie, other than getting a change of clothing and a knife, and he told me to go with Tex and do exactly what Tex told me to do."

"Did Tex tell you where you were going to go?"

"He told us that we were going to a house that used to belong to Terry Melcher."

"Was there a reason why you were going to that house?"

"The reason Charlie picked that house was to instill fear into Terry Melcher because Terry had given us his word on a few things and never came through with them."

"Did Tex tell you that he'd been at Terry Melcher's former residence?" Bugliosi asked.

"Yes. Tex said he knew the outline of the house because he and Charlie had been there talking to Terry."

"Did Tex tell you why you were going to Terry Melcher's former residence?"

"To get all of their money and to kill whoever was there."

"What happened when you arrived at the residence?"

For more than an hour Atkins testified, though less graphically, to the same account that she'd given Ronnie Howard, save one detail. "I remember seeing Sharon Tate struggling with the rope, and Tex told me to take care of her because Katie was asking for help with Abigail, and I saw Tex stab Abigail. When he came back in he told me to kill Sharon and I couldn't. In order to make a diversion, I grabbed her hand and held her arms and then I saw Tex stab her in the heart area. Then Sharon fell off the couch, to the floor, and all three of us went out the front door."

"Susan, how did you feel about what you had just done?" Bugliosi asked.

"I almost passed out. I felt as though I had just killed myself. I felt dead. I feel dead now."

"Susan, in what context would you and the other members of your family use the word *pig* . . . and *helter-skelter*?"

"Helter-skelter was to be the last war on the face of the earth. It would be all the wars that have ever been fought, built one on top of the other. Something that no man could conceive of in his imagination. You can't conceive of what it would be like to see every man judge himself and then take it out on every other man all over the face of the earth. And *pig* was a word used to describe the establishment."

An hour after the noon recess, Atkins's testimony was completed. Following two days of collaborating witnesses, the grand jury indicted Manson, Watson, Krenwinkel, Atkins, and Kasabian for seven counts of murder and one count of conspiracy to commit murder. Leslie Van Houten was indicted for two counts of murder and one count of conspiracy.

P.J.

TWO-NIGHT ORGY OF MURDER TOLD TO THE JURY BY TATE SUSPECT. I gazed at the newspaper headline and the accompanying pictures of the killers while my wife vented over the phone from Texas.

"You know, I've been hanging on for months, just thinking when they caught these assholes that I'd feel better," she cried. "But it hasn't made a damned bit of difference. All everyone keeps saying is, 'closure, closure, catching the killers should give you closure'—if I hear one more person say 'closure,' I'm going to strangle them. There is no goddamned closure!"

Even though I agreed, I tried to subdue her. "I know, honey, but at least there'll be some justice done here."

"How can you say that? Nothing a jury decides is going to bring our daughter home," she cried bitterly.

"All right, let's just calm down."

"Don't tell me to calm down. I'm madder than hell," she lashed out. "On top of everything else, the press is starting up again, writing about Sharon being part of a black and white racial war. What are they talking about? Sharon didn't have anything against Negros."

"They're not saying that," I patiently explained. "It has something to do with Manson ordering his people to kill Sharon. Then they tried to make it look like Negros did it to *start* a racial war. It's all nonsense, and you know better than to read the newspapers."

"Where are they getting this crap from?"

"I don't know. The DA is digging it all up from Manson's group. By the time you come home they'll have some answers."

"Have you made our reservations?"

"Yes, for the eleventh, after Manson's arraignment. By then things should calm down with all the media."

Throughout our conversation, my eyes had never left the newspaper photos of Sharon's killers. I may have been pacifying her on the phone, but my wife's anger paled in comparison to mine. I toyed with the pistol lying next to the paper. It was four o'clock. In forty-eight hours, Manson would arrive at the downtown Los Angeles jail.

MY PREDICTION THAT the media's interest would subside was premature. The funnel cloud of their storm caught the public's interest, driving them into a frenzy. Everyone wanted to see this monster who had glossed the headlines.

Jake circled the car around for the fourth time. "Jesus, what is this, the fucking witch hunts of Salem? There's not a parking spot within five miles of the place," he complained.

I was edgy. "Manson's supposed to be here in fifteen minutes. You mind waiting in the car?"

"I guess not—on one condition: You leave the heat behind."

I tried to fool him with a stoic look.

"Come on, I know you're packing, and it ain't the way to settle this score. You got a wife and kids coming home today. You do this, and they'll be fighting this shit alone. Now let's have it." He held his hand out for my gun.

I grudgingly laid the pistol on the seat. Before I shut the door he said, "And remember what Helder said: no closer than six feet to him."

I'd been forced to watch Atkins, Krenwinkel, Van Houten, and Kasabian being brought into Los Angeles via the television news, but I wasn't about to do the same with the man who'd supposedly masterminded Sharon's murder. With minutes to spare, I pushed through the crush of curiosity seekers. The double doors at the end of the corridor opened. A sudden surge by the crowd pushed me the last three feet to the barrier rope. Camera lights turned on from every direction as the entourage stepped through.

Ten deputies surrounded Manson; three were in the lead, two on either side of him, and three behind. In the center, Manson's head barely reached the shoulder height of the men escorting him. In spite of his cuffed hands, he walked with casual confidence, nodding and smiling at the reporters.

Camera flashes bounced off the walls and ceiling; their strobes caused a slow-motion effect. Over the past few months, my picture had appeared in the news. Given the chance, I was sure Manson would recognize me. When he was at arm's reach, I pulled down my sunglasses, making eye contact with the man who plotted and then ordered Sharon's murder. Manson's smile faltered as we locked eyes. It was a slow study of the enemy on both our parts.

Three feet beyond me, Manson twisted around, walking backward in order to continue the staring match. Pointing my finger first at Manson, then at my chest, I mouthed the words, "You're mine."

8

EVIL HAS ITS ALLURE

If the death penalty is to mean anything in the state of
California other than two empty words, this unquestionably
was a case for the imposition of the death penalty.

—VINCENT BUGLIOSI, MANSON TRIAL LEAD PROSECUTOR

Patti

The media swarmed the downtown courthouse like journalistic
sharks drawn into the feeding frenzy by the Manson Family. Months
before the trial started, reporters played a high-stakes, cutthroat game
for exclusive interviews with the suspects. Publication angles varied
from *Life*'s sinister portrayal of Manson with deranged eyes on their
cover to *Tuesday's Child* naming him "Man of the Year." The media's
focus on the Manson Family was an obscure blessing, as they'd all but
forgotten that we existed.

Information about the killers or their destiny meant little to me.
I was too young to understand justice or its system, so if "it" couldn't
magically bring my sister back to life, the outcome was inconsequen-
tial. On the other hand, my parents sought retribution and scanned
the headlines for the progress of the upcoming trial. SUSAN ATKINS
RECANTS HER GRAND JURY TESTIMONY. CHARLES WATSON FIGHTING

EXTRADITION; WILL HAVE A SEPARATE TRIAL. NEW PROSECUTION STAR
WITNESS, LINDA KASABIAN: IMMUNITY FOR HER TESTIMONY. UNITED
DEFENSE: MANSON AND FEMALE FOLLOWERS, ONE TRIAL, ONE DEFENSE—
NOT GUILTY PLEA EXPECTED. TRIAL OF THE CENTURY BEGINS TOMORROW.

P.J.

Manson followers camped out around the clock in front the Criminal
Courts building, preaching their leader's daily gospel. Their base, the
corner of Broadway and Temple, became an entanglement of supply
and demand for the curious. Hollywood tour buses added the spectacle to their route; it was the number-two attraction, second only to
the Hollywood Walk of Fame.

Nine floors above the faithful group's exhibition, Manson, Krenwinkel, Atkins, and Van Houten entered the stuffy, outdated Department 104, where fifty-two-year-old Judge Charles Older, an
ex–fighter pilot, brought his courtroom to order.

June 15, 1970, was a hot day. The portable air conditioners rattled
in protest against the windows. Considering the blizzard of pre-trial
motions from the defense, it should have been cooler for the first day
of the trial proceedings.

To accommodate the defendants and their attorneys, the bailiffs
cramped eight chairs around an L-shaped table. Clients and lawyers
alike formed a motley-looking crew.

Numerous defense attorneys came and went before the trial.
Manson handpicked the team that made the final cut for an obvious
reason. Irving Kanarek, Paul Fitzgerald, Daye Shinn, and Ronald
Hughes were the epitome of trial obstruction.

Alongside Vincent Bugliosi at the prosecutor's table sat co-counsel
Aaron Stovitz. Though lackadaisical in comparison to Bugliosi, Stovitz was an effective prosecutor and the stabilizing force when Vince's
agitation rose to the level of the overhead fans.

In a month's time, the defense and prosecution scrutinized 204
perspective jurors before agreeing on a twelve-member panel. On July

24, 1970, those jurors plus six alternates filed into the seats they would occupy for close to a year.

Spectators had lined up before dawn to secure a place at "the trial of the century." I was there early as well, waiting in a private corridor with the other witness scheduled to testify that morning. Two of us, who had never met, yet had much in common, sat quietly side by side on the bench, watching a bailiff walk down the hall. The bailiff stopped in front of us. "Mr. Tate?"

I nodded.

He then looked at the other man. "Mr. Parent?"

"Yes, sir."

"Gentlemen, opening statements have begun. Once completed, Mr. Tate, you will be the first witness called to the stand, followed by you, Mr. Parent. It shouldn't be long. I'm going back in. If anyone gives you trouble, press the button on the door. That will alert me, and I'll come right out," he explained.

I would have preferred to continue in silence, but Steve Parent's father extended his hand. "I'm Will."

"P.J.," I said, shaking his hand. Steve had been the fifth victim on August 9. It was his body that had been beneath the bloody sheet in the unknown car at Sharon's house. It was a clear case of being at the wrong place at the wrong time for the eighteen-year-old. He'd paid a late-night visit to the caretaker, Garretson, and encountered the killers on his way to the gate.

Will rested his head against the wall. "Rumor has it in the papers that you want to kill Manson. Wish I had the guts to do it myself."

"Don't believe everything you read."

"How's your family holding up?"

"How the hell do you think we're holding up?" I said bitterly.

"I'm sorry. I didn't mean to pry. I just thought . . . well . . . we're not doing so hot." Parent's voice cracked.

I softened. "Knee-jerk reaction. It's been a long time since anyone asked me that question with sincerity." I offered him a cigar. "For the record, we're doing shitty."

Will accepted a light, then said, "It's been almost a year, but my wife still won't talk about it. Every time she hears a car pull in the driveway, she perks up, like it's Steve coming home from work or something. She won't let me touch anything in his room. His pajamas are still lying on the bed."

"Yep, same with mine. She spends most of the time in denial. You won't catch her within five miles of this courthouse because she can't bear to hear the slightest detail of what happened."

"I guess in time they'll open up," Will said.

"Probably," I commented, though I didn't believe it. Time was disintegrating my marriage. Nightly, the gentle shaking of the bed awakened me to find my wife crying. If I tried to comfort her, she silently turned away.

We'd been married twenty-seven years. I'd known her a few years longer than that, since high school actually, except I didn't have the courage to ask her out then. But when I saw her again in 1942, I had a few more years' experience and a few more drinks under my belt, which encouraged me to ask her for a dance. Artie Shaw and Glenn Miller played beneath the Sylvan Beach Pavilion that overlooked Galveston Bay. Texans like to two-step, but they also like to dance hand in hand while the big bands play. And from that first dance with Doris, I was a man smitten.

Three wars and my daughter's murder couldn't change my feelings; I was still smitten. But with each sunset, a couple of bricks were added to the wall building between us, and I didn't know how to stop the construction. Will Parent interrupted my thoughts. "What do you think is happening in there?"

"With any luck, Bugliosi's inspiring the folks on the jury to set up the gas chamber."

VINCE BUGLIOSI TUGGED at the vest of his three-piece suit, smoothing it to perfection. "In this trial," he said to the jury, "we will offer evidence of Manson's motives for ordering these seven murders. Besides the motive of Manson's passion for violent death and his extreme anti-establishment state of mind, the evidence at this trial will show that there was a further motive which was almost as bizarre as the murders themselves.

"Very briefly, the evidence will show Manson's fanatical obsession with 'Helter Skelter,' a term he got from the English musical recording group the Beatles. Manson was an avid follower of the Beatles and believed that they were speaking to him through the lyrics of their songs.

"To Manson, *Helter Skelter* meant the black man rising up against the white establishment and murdering the entire white race, that is, with the exception of Manson and his chosen followers, who intended to escape from Helter Skelter by going to the desert and living in a bottomless pit, a place that Manson derived from Revelation 9, the last book of the New Testament.

"The evidence will show that although Manson hated black people, he also hated the white establishment, whom he called 'pigs.'

"The evidence will show that one of Manson's principal motives for the Tate/LaBianca murders was to ignite Helter Skelter. In other words, to start the black-white revolution by making it look like the black people had murdered the five Tate victims and Mr. and Mrs. La-Bianca. Thereby causing the white community to turn against the black man, and ultimately lead to a civil war between blacks and whites, a war Manson foresaw the black man winning.

"Manson envisioned that the black people, once they destroyed the white race and assumed the reins of power, would be unable to handle the reins because of inexperience and would have to turn over the reins to those white people who had escaped from Helter Skelter. That is, turn over the reins to Manson and his followers."

Four hours later, the bailiff emerged from the courtroom to find Will and me as he'd left us. "Mr. Tate, they're ready for you. Step up to the table here and empty your pockets."

Expecting the search, I carried only three items: my wallet, a lighter, and a cigar pouch. "Place your hands on the table," the deputy said, as he began to pat me down.

"Careful down there," I nervously joked.

He remained stone-faced. "Sorry, we're searching everyone, and they told me to give you a real thorough check. Okay, let's go." He paused at the door. "Let's not make any headlines today. Don't do anything crazy in there. Just try to concentrate on the questions the lawyers ask you. Are you ready?"

I pulled back my shoulders, took a deep breath, and gave a nod. All the way to the witness chair, I concentrated on the deputy's back like a side-blinded horse.

Right arm raised, I faced the clerk and was sworn in. "Please be seated, Mr. Tate," Judge Older said.

I kept my look downcast, avoiding Sharon's killers sitting less than twenty feet from my grasp. The frantic scratching of the reporters' pencils cavernously echoed amid hushed anticipation. Equally loud, seconds clicked away on the wall clock. Simple testimony, I thought. Identify Sharon, Woytek, Gibbie, and Jay, then get the hell out of Dodge.

My hands white-knuckled the armrests of the chair. What in the hell was taking the prosecutor so long?

Temptation won out. I looked up. My eyes locked onto Manson with his perpetual dumb-ass grin. I had read about his so-called hypnotic stare that he tried to use to intimidate people. But that crap wasn't going to wash with me. I gave him my own little grin and stared back with a look that said I'd as gladly snap your neck as I would testify. It didn't take long before his smile disappeared and he looked away.

My inspection drifted down the defense table to the women. They were so young, giggling with the innocence of teenagers at a slumber party. On closer examination, their facial expressions emitted complete dispassion. Noticing my scrutiny, Atkins seductively licked her lips while her finger tapped on a picture of Sharon that lay atop the files.

Her action was all the goading I needed. The deputy who searched me for hidden weapons didn't understand that the most lethal weapons I had were in plain sight: my mind and body. I counted the armed personnel and their positions; figuring the odds of whether I could jump the witness stand to the defense table before they stopped me. Inside five seconds, I could have one of their necks, but I'd have to make a choice: Atkins, the knife wielder, or Manson, the instigator. Aaron Stovitz moved to the front area, blocking my view. "Would you please state and spell your name, sir?"

Before I could respond, Irving Kanarek sluggishly lifted his heavily framed body from the chair. "I make a motion to exclude the witness, Your Honor."

"Denied. Sit down, Mr. Kanarek," Older admonished Manson's lawyer, and then turned to me. "You may answer the question."

"Paul James Tate, T-A-T-E." . . .

"What is your business or occupation, sir?" . . .

"I was an intelligence officer . . . as lieutenant colonel." . . .

Stovitz handed me a photo. "I show you Exhibit 1 for identification, Sir. Do you recognize the person depicted in that photo?"

I glanced at the photo and then looked at Manson. "Yes, I do. That is my daughter, Sharon Tate." . . .

"And how old was Sharon Tate when you last saw her?" Stovitz asked.

"She would have been twenty-six." . . .

"And when was it that you last saw her in life, sir?

"It was in July, actually, the day of the moon landing in 1969." . . .

"And how did you know it was the day of the moon landing?"

"Well, all of us watched the landing on television as everyone else was doing."

"And when you say 'everyone else,' who else was there, sir?"

"Well, at the time there was Abby [Gibby] Folger and Woytek Frykowski. Later Jay Sebring came in from San Francisco and came directly from the airport and joined us . . ."

"And where was this taking place, sir?"

"At Sharon's house." . . .

"Now, directing your attention to Exhibit 2 for identification. You mentioned a person by the name of Jay Sebring. Is that the person shown in Exhibit 2 for identification?"

"Yes, it is. That is Jay Sebring." . . .

"And . . . the persons depicted in Exhibit 3 for identification . . . ?

"That is Woytek Frykowski . . . and Miss Abigail Folger." . . .

"Now, on this day in July . . . did you just visit for a few hours or what?" . . .

"My wife and my other two daughters and myself spent the entire day . . . and left there at 11 or 12 o'clock that night."

"When was the next time you either spoke to your daughter or saw her again after that?"

"Of course I never did see her again after that." . . .

"Your Honor, that's all we have for this witness," Stovitz told the court.

I started to leave. "Hold on, Mr. Tate," the judge said, and then asked the defense if they wanted to cross-examine me.

Atkins' lawyer, Daye Shinn, stood. "I have a few questions, Your Honor."

The Asian attorney spoke as smoothly as the used car salesman he once was; overly friendly as he trapped the unsuspecting into a raw deal. "Mr. Tate, you stated that you knew Mr. Sebring since 1964?"

"Yes, I did, approximately 1964."

"Did you know him socially?"

"Socially? I had been to the opening of his shop. I suppose you could call that socially."

"And did you have any business contacts with Mr. Sebring?"

"No. I don't know exactly what you mean by business contacts. In what respect?"

"Well," Shinn said dramatically, "in your association—"

"I was in the military during this period, not the hair business."

"You did see him from time to time?"

I looked quizzically toward the prosecutors. "Yes, from time to time." . . .

"Did you visit his house?" Shinn questioned.

"Yes, yes I have been to his house."

The volume of the lawyer's voice pitched a decibel higher. "And you met him at your daughter's house, too, at times?"

"Yes, of course."

Shinn turned toward the jury and roared, "And did you attend any parties at your daughter's house or Mr. Sebring's house?"

Shinn's destination was clear, victim defamation. I'd had no choice but to sit back and watch it happen in the press, but I wasn't about to let it happen again in a court of justice. Two could play this distraction game. "Your Honor, do you suppose he's asking me or the jury?"

Shinn wheeled. "What? I—" he stammered. "I'm asking you, of course."

"Oh. In that case, no, I did not," I lied. . . .

Caught off guard, Shinn floundered with less steam. "As to Mr. Frykowski, . . . how long had you known him?"

"I would say roughly about four months . . . I had not known Mr. Frykowski long." . . .

"And how about Miss Folger?"

"The same, the same," I snapped . . .

"During your association with these various people I just mentioned, . . . did you at any time observe them under the influence of either drugs or alcohol?"

I folded my arms in defiance. "Never . . . You got any snake oil to sell with this load of manure?"

"Objection, Your Honor," Shinn whined at the same time that Stovitz objected to the question being immaterial.

"Sustained. The answer will be stricken. The jury is admonished to disregard it. . . . Continue."

I was confused on which objection had been sustained so I asked, "Well, is his question about drugs or about alcohol?"

"Objection, Your Honor," Stovitz intervened, "immaterial."

"Sustained. The answer will be stricken . . . I have to agree. Mr. Shinn." . . .

Shin asked to approach the bench. When he did, Bugliosi, Stovitz, and the rest of the defense lawyers huddled around the judge. I caught just bits and pieces of the argument, but I believe Shinn wanted to continue on his alternate drug motive quest and introduce evidence that tied Jay not only to the drug dealer Billy Doyle, but also to the whipping Doyle supposedly received by Jay and Woytek. Thankfully, the judge refused to allow that line of questioning and told Shinn he would be on a very tight leash for the remainder of my cross-examination.

When court resumed, Older said, "Let's proceed, Mr. Shinn."

"Mr. Tate, did you ever see Jay Sebring use a whip on someone in a sexual—"

Judge Older's hand cracked down on the bench. "That question is stricken . . . Your examination is over, Mr. Shinn. . . . You are excused, Mr. Tate, subject to being recalled . . . Please let the court know if you plan to leave town."

"That will be fine, Judge. Starting tomorrow, you can find me right there," I pointed to the gallery, "in one of those butt-numbing seats."

On my way out, I passed the seasoned court reporters, such as Theo Wilson and Linda Deutsch. In the corridor, I found the less professional version. Their cameras flashed, their questions stung.

"Is it true that Sharon and the others used Manson as their drug connection?"

"Did Manson really attend orgies with Sharon at Cielo Drive?"

I pushed passed them. "Kiss my ass."

The following Monday morning, I sat in a seat with the best vantage point, closest to the wall, behind the prosecution, and to the side of the defense table; all of which provided an unobstructed view of the jury—not for my benefit, but for theirs. Whether the jury would admit it or not, seeing my face, front and center, day in and day out would weigh on their mind for a conviction.

Theo Wilson laughed behind me. "Oh, God, what next?"

Atkins, Krenwinkel, and Van Houten entered, wearing blue satin capes and mopping blood that still dripped from the X they'd carved into their foreheads.

Manson saved his theatrics for the jury. The instant they were seated, he held up a copy of the Constitution, then dramatically dumped it into the waste can.

Vince Bugliosi, grown so accustomed to the defendant's antics, barely noticed what the columnists jotted down for their afternoon report. He looked up from his notes. "The People call Linda Kasabian."

A wilted blossom of flower power, Kasabian entered the court with her head tilted as if her mind weighed heavily. Near the defense table, her dispirited eyes remained downcast, avoiding her co-defendants, who rapidly ran their fingers over their lips, signaling "blabbermouth."

For the moment, Kasabian remained a defendant, held without bail, and charged with seven counts of murder. However, upon completion of her testimony, the state of California planned to grant her immunity.

"Raise your right hand and repeat after me," the clerk said.

Kasabian tucked back her blond locks, revealing a pale, gaunt face. Before she had a chance to raise her hand, Irving Kanarek gave a jolt. "Objection! On the grounds that this witness is not competent and is insane!"

Bugliosi shot up. "Wait a minute! Your Honor, I move to strike that, and I ask the court to find him in contempt for gross misconduct. This is unbelievable on his part."

The lawyers' exchange set the tone for Kasabian's eighteen days on the witness stand. Kanarek maniacally objected to almost every question the prosecutor asked, often finding numerous, irrelevant objections for the same question. Coinciding with Kanarek's misbehavior was a flow of outbursts from the defendants, that is, until Bugliosi led everyone back to August 9, 1969.

Kasabian's testimony of murder dropped a blanket that overlay the spectators' rustling of papers, clearing of throats, and bodies shifting uncomfortably on the benches. Closer to the witness stand, Atkins, Krenwinkel, and Van Houten stopped giggling. Manson concentrated on doodling. The defense attorneys halted their objections. The jury edged forward in their seats.

"What happened after you, Katie, Tex, and Sadie walked up the hill to the Cielo gate?" Bugliosi asked.

"We climbed over a fence, and then a light started coming toward us. Tex told us to hide in the bushes. A car pulled up in front of us, and Tex leaped forward with a gun in his hand and stuck his hand with the gun at the man's head. And the man in the car said, 'Please don't hurt me, I won't say anything.' And Tex shot him four times."

"After Tex shot the driver, what happened?"

"The man slumped over, and then Tex put his head in the car and turned the ignition off. . . . We all proceeded toward the house, and Tex told me to go in back to see if there were open windows or doors, which I did."

"Did you find any doors or windows open?

"No. . . . When I came around from the back, Tex was standing at a window, cutting the screen, and he told me to go back to the car and wait. I waited for a few minutes at the car, and then all of the sudden I heard people screaming, 'No, please, no.' It was just horrible. Even my emotions cannot tell you how terrible it was. I heard a man scream out, 'No, no.' After that, I just heard screams. I don't have any words to describe those screams. It was just unbelievably horribly, terrible."

"Were the screams loud screams or soft screams, or what?" Bugliosi asked.

"Loud. Loud," Kasabian said, as if yelling over voices in her head.

"Did the people appear to be pleading for their lives?"

"Yes."

"What did you do when you heard these screams?"

"I started to run toward the house."

"Linda, what happened after you ran toward the house?"

"There was a man just coming out of the door, and he had blood all over his face, and then he leaned against a post. We looked into each other's eyes for a minute. Then he just fell into the bushes. Then Sadie came running out of the house, and I said, 'Sadie, please make it stop.' And she said, 'It's too late.' And then she told me that she couldn't find her knife. And while this was going on, the man had gotten up and I saw Tex on top of him, hitting him on the head and stabbing him, and the man was struggling. Then I saw Katie in the background with the girl, chasing after her with an upraised knife."

"Katie was chasing someone?"

"Yes. A woman in a white gown."

"When Tex was stabbing this man, was the man screaming and struggling?"

"Yes."

"How many times did Tex stab this man?"

"I don't know. He just kept doing it and doing it and doing it."

"Do you know what he was screaming?"

"It was beyond words, it was just screams."

"What did you do next?"

"I just turned and ran to the car down at the bottom of the hill."

"Did Tex, Sadie, and Katie eventually come to the car?"

"Yes."

"What happened then?"

"I had started the car, and Tex came over and told me to turn the car off and to push over, and he seemed really uptight, because I had run to the car."

"Did Tex then drive off?"

"Yes."

"Did Katie and Sadie say anything as you were driving off from the residence?"

"Yes. They complained about their heads, that the people were pulling their hair, and that their heads hurt. Sadie said that when she was struggling with a big man, that he hit her in the head. Also, Katie complained that her hand hurt."

"Did she say why her hand hurt?"

"She said when she stabbed that there were bones in the way and she couldn't get the knife through all the way, and that it took too much energy. I don't know her exact words, but it hurt her hand."

"Did Katie say anything about one of the girls inside the residence?"

"Yes, she did. She said that one of the girls was crying for her mother and for God."

"Would this be a good breaking point, Mr. Bugliosi?" Judge Older asked.

"Yes, Your Honor."

"Then we'll adjourn until nine o'clock tomorrow morning."

DORIS AND PATTI were finishing dessert when I got home.

"We gave up on you," Doris said. "I wish you'd call if you're going to be this late."

"Sorry," I halfheartedly told her.

"Sorry? The prosecutor and the judge have full-time bodyguards because of these maniacs. You waltz around without protection, and I don't know if you're dead or alive."

"I said I was sorry. It was a long day in court. Today was Linda Kasabian's testimony of the murders. That girl's a double-edged sword; she could have saved them."

"Not in front of Patti."

"Why the hell not?" My temper flipped. "Patti, you know what's going on in your sister's trial, don't you?" I wasn't looking for an answer, nor did I wait for one. "Everyone knows what's happening in the goddamned trial except for you!"

"What would you have me do? Sit there every day, tortured with the details of how my daughter took her last breaths? Would that make you feel better, P.J.?" she cried.

"At least everyone would know she had a mother!" I stormed out of the room.

"You bastard," she chased after me. "I don't need to be there to prove I love Sharon. It's in God's hands to take care of those monsters."

"What God?" I bellowed. "The same God that watches over our children? If it is, you better keep your eye on that little one in the kitchen, because *there is no God!*"

"That's a cop-out." Her voice calmed. "Evil's alluring, isn't it? It's an easy way out, the exact same escape that lured Manson's followers. Some days I may hide my head in the sand, but at least I'm not hiding behind hatred."

Later, I eavesdropped at the bedroom door I was locked out of. "Are you and Dad going to get a divorce?" Patti asked Doris.

"No, honey."

"Then how come you have separate bedrooms?"

"Your daddy and I are just on different paths right now. We'll find our way back sooner or later."

I went to my new sleeping quarters, unsure if she was right.

THE REMAINDER OF the Manson trial was an outlandish experience with something for everyone. Hippies, movie stars, and tourists piled into the courtroom to witness the spectacle. Tattooed bikers, cowboy stuntmen, hookers, Hollywood agents, music producers, and the prosecution's star witness Yana the Witch, testified about "Elmer" the pot plant, acid trips, karma, the Bible, the Beatles, sex orgies, bisexuality, war, love, and murder. It was a trial presided over by a judge who carried a .38 revolver under his robe. And a trial where oddities of exceptional circumstances became the norm.

A woman claiming to be the Whore of Babylon stopped the proceedings when she screamed, "I have proof that key prosecution witnesses were coerced, bribed, and threatened!"

As if to provide corroborating evidence to the prosecution's theory that Manson was the puppet master of his co-defendants, the female defendants followed his every lead. If he turned his chair back to the judge, they followed suit. When he stood with his arms outstretched in the form of the crucifixion, they stood to do the same. One day, he simply nodded at them and they all rose, lifted their skirts, and then bent over the table to moon the audience—thankfully, they wore underwear.

Disguised, Manson Family member Squeaky Fromme slipped into court one day only to be ejected after she jumped on my back. Smacking the top of my head, she yelled, "You should know better than anyone that we were just at war! Who are you judging? Follow your own reflection, the guilt you find is yours!"

Deputies dragged her out to the corridor, where her scream was heard all the way to the elevator. "It's not our fault, they brought it on themselves! You've all brought judgment down on yourselves! Prepare for the end!"

No one bothered to ask if I wanted to file charges against Fromme—it was just another day in this out-of-control trial.

That interaction with Fromme wasn't my only personal encounter with the Manson Family. On another day, I entered the courthouse men's room still thinking about Virginia Graham's testimony. She recounted Susan Atkins's confession in which Atkins provided details of how she'd killed Sharon. I'd always believed that my genetic disposition made it impossible for me to ever raise a hand to a woman. But Atkins was doing a great job challenging that belief. Given the opportunity, I wouldn't have hesitated to make her suffer tenfold for what she did to my baby.

I zipped up with that thought and then noticed for the first time that the men on either side of me were focused more on me than the urinal. A third man exited a stall and left without washing. Although the dark-haired one to my left fingered a sheathed knife, I was fairly confident that both men were there for intimidation purposes only. Nevertheless, I squatted as if to tie my shoe and then I charged him. By the time we slammed into the far wall, three deputies were on us; one yanked my arm behind me and jerked it upward to restrain me. When he spun me around, I saw that the second man had left. As with the Fromme incident, the deputies weren't interested in the details of the latest sideshow of the circus. Within the hour, I was ejected from the courthouse for the day.

Weeks later, I saw the two men on the evening news. LAPD had issued warrants for the arrests of Bruce Davis and Steve Grogan, under suspicion for the murder of Donald O'Shea, a ranch hand at Spahn's Movie Ranch.

October 5, 1970, was a particularly bad day.

When Vince Bugliosi finished his direct examination of Sgt. Whiteley, Manson raised his hand. "May I examine him, Your Honor?"

"No, you may not," the judge said.

"Are you going to use this courtroom to kill me? Do you want me dead? The minute I see that you are going to kill me, you know what I'm going to do?"

Older looked at him curiously. "What are you going to do?"

Manson smiled. "You know, you've studied your books. Do you know who you are talking to?"

"If you don't stop, Mr. Manson, and I order you to stop now, I will have to have you removed as I did the other day," Older threatened.

"I will have you removed if you don't stop. I have a little system of my own," Manson challenged.

"Fine. Let's move on. Mr. Bugliosi, call your next witness."

"Do you think I'm kidding?" Manson's voice boomed. In a rage of adrenaline and with a sharpened pencil, he leaped across the defense table toward the judge. A deputy tackled Manson before he hit his destination. As six deputies carried him out, he screamed at Older, "In the name of Christian justice, someone should cut your head off!"

The female defendants joined in. "You're just a woman," they chanted in unison. "You are just a woman, that is all."

"Remove the defendants!" Older barked over them.

Exciting as the outburst was, the cliff-hanger of the proceedings was just around the corner.

The end of the trial neared. Closing arguments were scheduled for Monday, November 30, 1970. Instead, the courtroom occupants impatiently waited for Leslie Van Houten's attorney, Ron Hughes, to arrive.

At ten o'clock police knocked on Hughes's front door. No one answered.

By the end of the day, the only thing the police discovered was that Hughes had gone camping in the Sespe Hot Springs the weekend before. Beyond that, it was anybody's guess, and everyone had the same one: the Manson Family had killed him.

Judge Older assigned Maxwell Keith as co-counsel for Van Houten, which delayed the proceedings until December 21.

The morning the trial resumed, Leslie Van Houten stood to announce, "I had nothing to do with Ronald Hughes's disappearance, Judge Older. I'm beginning to wonder what *you* did with him."

"Miss Van Houten, sit down, or you will be removed."

"You stand up!" Van Houten retorted. "Hey! Look at me when I'm talking to you. I'll take care of you, you'll see!"

"Remove the defendant."

A deputy grabbed her arm. "Get off of me, you fucking bitch!" Van Houten yelled, then punched her.

"Where's the justice now!" Atkins joined in by taking a swing at another deputy. "Who are you on judgment day?"

Judge Older banged his gavel. "Remove *that* defendant!"

Manson leaned back on his chair, singing, "The old gray mare ain't what she used to be. She's a judge now."

Older remained placid. "Mr. Manson, you are now disrupting this trial."

"That old black magic that you keep so well," Manson sang.

"I order you once again to stop this, sir."

"You have been ordering me forever," Manson stated. "All my life you have ordered me. You charge me with murder, and you say I have rights, and you hold up the rights in front of me, but you give me none."

"Mr. Manson, this will be your last warning. Mr. Bugliosi, let's proceed."

"You are going ahead, but are you going ahead to look at yourselves?" Manson turned toward the jurors. "Look at all of you. Where are you going? You're going to destruction, that's where you're going. You will end up being judged. It's your judgment day, not mine. I've already judged me."

With an outstretched arm, Krenwinkel yelled, "Hail Caesar!"

"That's it! Remove all the defendants. It is perfectly clear that this was a calculated performance by the four of them to interrupt the proceedings, and they will not be brought back into this courtroom again during the remainder of the guilt phase of the trial. Let's proceed, Mr. Bugliosi."

Vince collected his papers in the quieted arena and moved to the podium. It was so quiet, in fact, that the prosecutor startled everyone when he loudly declared, "It was your client, Mr. Kanarek, who ordered the commission of these horrible murders! These men and women who followed him were not suffering from any diminished mental capacity. They suffered from a diminished heart, and a diminished soul. . . . There is a penetrating spotlight on those two dark nights of murder. Charlie Manson knows that a violent death and brutal murder is the ultimate wrong, or he wouldn't be fighting for his life now. Charles Manson is on trial because he is a cold-blooded, diabolical murderer. As sure as I'm standing here, as sure as night follows day, these defendants are guilty."

Two days later, Vince closed his notebook and his final summation with a last thought. "Sharon Tate, Abigail Folger, Jay Sebring, Woytek Frykowski, Steven Parent, Leno LaBianca, and Rosemary LaBianca are not here in this courtroom now, but from their graves, they cry out for justice."

SUITCASES FILLED THE trunk of the car for a long weekend at Mammoth Mountain. On our way out the door, the phone rang.

"Leave it, P.J.," Doris said.

I hesitated. "The jury might be in."

"It's our anniversary. They're going to make their decision whether we're there or not. Vince has the number at the cabin."

I stubbornly went to answer the phone, and my equally stubborn wife headed for the car. "Let's go, Patti," she said.

"What about Dad?"

"Just get in the car."

A minute later, I was back at the car. "They've reached a verdict."

"I don't care, we've planned this trip for months," she complained. "We'll find out the verdicts on the news."

"You take the girls and go. I'll meet you up there tonight." I paused, suddenly confused. "Where's Debbie?"

"Where's Debbie?" She laughed hostilely. "That's just it, you don't have a clue what's going on with your family. You've completely tossed us aside for that trial. I'll bet you didn't even remember yesterday was Sharon's birthday."

"The hell I didn't. Who do you think put the flowers on her headstone? Oh, wait, I forgot, you've never seen her headstone!"

She smacked me. "Stop it!"

"When are you going to wake up? *That trial* is because they killed Sharon. She's not off doing a movie somewhere; she's never coming home. And I'm going to make damned sure those sons of bitches never go home to their parents, either. Now I'm going down to that courthouse to see that commitment through."

We both got into our cars and went our separate ways.

DEFENDANTS AND PROSECUTORS enter a trial equally confident that they will win the game. The ensuing weeks of strategic testimony is a pressure cooker of emotions for anyone with a personal stake in the outcome. The moments before a verdict is read are riveting seconds of saturated tension masked by the calm of the poker-faced lawyers, jurors, and defendants.

The unpredictability of the reactions to the verdict by the accused or the victim keeps the bailiffs and deputies alert to confrontation. Considering the track record of the Manson trial, they anticipated the worst.

A dozen deputies lined the railing that separated the defendants from the spectators. I sat, quietly focused on the killers' glib entrance; however, their nonchalance contradicted the apprehension of their wandering glances. Once they were seated, Older said, "The clerk will read the verdicts."

"We the jury in the above-entitled action, find the defendant, Charles Manson, guilty of the crime of murder of Abigail Folger, a human being, in violation of section 187, Penal Code of California, a felony as charged in Count 1 of the Indictment, and we further find it to be murder in the first degree."

In all, there were twenty-eight verdicts read, one for each victim and one for each murderer. All guilty, all murder in the first degree.

With the final verdict handed down, Manson was the first to break the silence. He looked to the jury. "You're all guilty." Then to Older. "We are still not allowed to put on a defense?"

The judge ignored Manson. He banged his gavel. "Court is adjourned."

Manson fought against the restraining deputies. "You won't outlive that, old man!"

The girls lost their trial-long, happy-go-lucky conduct. "Your whole system is a game," Van Houten seethed. "You blind, stupid people, your children will turn against you!"

The deputies pushed the killers toward the lockup. Atkins broke free, darting toward the gallery. "Better lock your doors and watch your own kids!" she managed, before deputies wrestled her to the ground.

As quickly as the uproar began, it ended with the doors closing behind the prisoners, and the end of a more trusting era.

9

LIFE GOES ON. OR DOES IT?

It's hard for me sometimes to accept the fact that people choose
to believe that I absolutely cannot change. That I was something
at nineteen, and what I am at thirty-three is irrelevant because the
life of the one they loved ended when I was nineteen. And though
I understand it, it's very difficult because life goes on. And I go on.

—LESLIE VAN HOUTEN

Patti

Countless times throughout my childhood, I'd heard Dad's caution-
ary advice: "Never assume. It will bite you in the ass every time."

Like many kids, I ignored my parents' advice. I assumed that the
guilty verdict and death sentence of Sharon's killers would press a
magic button that'd cause them to disappear as quickly as Sharon
had. I assumed the close of the trial was the light at the end of the
dank and gloomy tunnel that held me hostage to a night of intangible
evil.

A year later, I was still massaging my wounded rear end as
the Manson Family and their heinous crimes continued to gain
infamy—as the media's ever-evolving fascination with them fed the
steady diets of the inquisitive.

The group's cases filled the dockets of the Los Angeles courts for years after the main trial's conclusion.

California extradited Charles Watson from Texas and convicted him with a death sentence for the Cielo and Waverly Drive murders.

In separate trials, Manson, Bruce Davis, and Steve Grogan were each found guilty for killing Spahn's ranch hand Donald O'Shea.

Gary Hinman's parents received justice for their son's murder when a jury convicted Beausoleil, Davis, Manson, and Atkins.

Toward the end of 1971, after an intense gun battle, the LAPD apprehended Manson followers in the midst of a weapons shop robbery. The shootout made the six o'clock news. I nervously watched the screen as the police wrestled the crew into custody; positive that the killers were closing in because the event took place just ten minutes from our house.

The trial for the armed robbers began as the Superior Court reviewed Atkins, Manson, Van Houten, and Krenwinkel's appeals to overturn their convictions. In the end, only Van Houten received a new trial due to defense attorney Ronald Hughes's untimely disappearance. The jury in her succeeding trial reached the same guilty verdict.

The sheriffs eventually recovered Hughes's body near the Sespe Hot Springs, but they failed to find enough evidence to warrant an arrest for murder.

Coinciding with the trials, publishing houses turned out books on Manson and the murders faster than soap opera scripts were written. Over the next few years, Sharon's case remained a top seller for the tabloid papers and movie magazines. Television shows referenced the murders, news programs ran special commentaries, and *Manson,* a feature-length film, was released nationwide.

The aftermath of the trial extinguished the optimistic light of escaping the tunnel. For the time being, I was still a prisoner of August 1969, where Sharon and Manson's faces were hauntingly intertwined, in a clash of love and hatred.

Although the fall of 1971 promised to be an exciting time as I entered my first year of high school, it proved instead to be a continuum of sleepless nights, hazy days, and insecurity; all deliberately hidden behind a smile and a lighthearted disposition that I rehearsed each morning on the walk to school.

I made every effort to avoid sympathetic gazes or questions about my well-being as they only triggered memories of Sharon's murder. I eventually made friends. I kept my grades at a C average, hung out at the beach after school, went to football games on Friday nights, and even made the ballot for homecoming princess; no one could have guessed that I was falling apart, harboring an intimate fear that blended neatly with my depression.

For a short period, I tried finding late-night reassurance by nestling beneath the down blanket of Mom's bed. Protected by her warmth, three door locks, bars on the windows, and a loaded gun in her armoire, it seemed a safer bet. We lay awake night after night, side by side, silently watching the walls shimmer from the Moon's reflection off the rippling swimming pool. Some nights the Santa Ana winds whistled through the trees; other times, they howled as if to instill in me even more fear of the obscure elements of this world.

Slowly, I withdrew from the nightly safeguard of my mother's bed because it ultimately became a troubling reminder of why I needed the security of her arms in the first place. Back in my own room, the radio's steadfast signal kept me company through the long nights, reaffirming that life *could* triumph through the witching hours.

After everyone else went to bed, I routinely crept through the house, inspecting the door and window locks. Then I headed into the kitchen to confirm that all the knives were safely concealed in the drawers.

In bed and watchful till dawn, I regularly checked the phone for a dial tone and peeked through the curtains to ensure it was the breeze rustling the bushes at my window and nothing more.

During those lonesome hours, it didn't matter how hard I tried to direct my thoughts toward the golden memories of Sharon. The instant my eyes closed, nightmarish images fired at me, obliterating any fleeting brightness I'd managed to create.

I'd dream of row upon row of ghoulish corpses lining the walls of a morgue. Or delight in seeing Jay's car in our driveway, until a look through the tinted windows revealed a flaming body, melting into the driver's seat. I'd see Sharon off in the distance, angelically floating toward me in a white minidress. As she'd draw nearer, webs of blood would spore across the fabric, dying it crimson. And then there was Winnie Chapman pounding on our window as Mom and I ate breakfast, my mother completely apathetic to her voice, shrieking, "Mrs. Tate! You've got to help! They're dying! They're bleeding all over the place!"

Within the nightmare, I awakened in a bathtub brimming with blood.

The killers manifested wickedly in those horrific dreams.

Watson, dressed in black, with a starkly white skull, an elongated jaw, and protruding razor-sharp teeth, would tower over me. A black cowboy hat resting atop his skinless head. Like a Texas gentleman, he'd tip the hat forward just before striking. Susan Atkins would materialize with long black hair that cloaked her head like the hood worn by the shadow of death, her face shrouded with the exception of two fiery, pinpoint eyes.

Stirred by Sharon's screams piercing my consciousness, I'd awaken with drowning eyes, unable to focus as I crossed back to reality. Curled up from the gnawing ache of terror commingling with sorrow and guilt, I'd plead for God to reverse time's hand. Convinced that if I'd stayed with Sharon I could have saved her—or at least perished with her; it seemed a better alternative to surviving.

By the time I'd secretly reached into my mother's medicine cabinet for her bottle of Valium, I was desperate to relinquish the predators who vengefully cornered my sanity.

I held out until midnight before slipping the pill between my lips. Although I was nervous and curious about what would come, I had barely lingered on the mystery before my body tingled from the dissolved remedy coursing through my system.

The narcotic warmly caressed me as I rode its wave higher and higher, enjoying the calm, yet afraid to fully release the hold on my domain. I forced my eyes to rest, and recalled a line from the movie Roman was making based on Shakespeare's *Macbeth*: *The innocent sleep. Sleep that knits up the ravell'd sleave of care. The death of each day's life.* My hands unclenched, my pulse slowed its hammer, and then I slept without demons.

The dream was strikingly bright with colors blooming through brushed edges. As if the long-awaited response to my prayers was at hand. I find myself in the Cielo living room. Sharon and I sit at the piano, the keys of which hold the imprint of the artists whose fingers had pulsated across the ivory in the past. John Lennon, Jim Morrison, John Phillips, Neil Young, and a host of others whose music had permeated my sister's home. Sharon takes my hand and repeatedly guides me through the five simplistic notes to John Phillips's song *Safe in My Garden*. We pause, and then she says, "There's nothing to fear from the past. In time, it will bring you freedom, but for now, I want you to let it go."

My eyebrows furrow. "No!"

Her fingertips smooth the creases on my forehead. "*You must.* When you're ready to look back, I'll be waiting." Her hand reaches to close my eyelids.

When they reopened, my bedroom was shining with the light of another day. I lay there motionless, reveling in the sentimentality of the night's passing, eager to return to the medicine chest for another trip.

In the years to come I experimented with every drug I could get my hands on, searching for the halcyon's wings to lift me above the stormy horizon of my mind and into obliterating clouds, until yesterday would become an indistinct notion.

THE STEAMING WATER from the showerhead cascaded over my body in the chill of February's eighteenth morning. It was a new year, and I felt like a new person with the residual of the previous night's drug, MDA, still massaging my brain into placidity.

Twelve hours earlier, my friends and I lay next to the surf under the spell of the drug in unrestrained euphoria, hoping to reach beyond the stars, into the perimeters, and on to the mystical part of the universe where Jim Morrison promised we'd find freedom.

The bathroom radio played Tommy James's *Draggin' the Line* as I rhythmically kneaded the shampoo into a playful lather. I was in love, and though my prince didn't know of my existence yet, I plotted how I'd win his heart later that day at the beach. *Jeans or skirt? Hair loose or tied back?* I leaned in to kiss the warm water, daydreaming of how his lips would feel when they fluttered against my skin for the first time.

A newscaster punched through the airwaves, his voice echoing off the tiles, disrupting my romantic trance. "Welcome to the ten o'clock news. Today, prisoners on California's death row breathed a sigh of relief when they learned of the state's decision to abolish the death penalty.

"The Supreme Court's ruling cites the reason for the abolishment as a practice that degrades and dehumanizes all who participate in the process. The high court also noted that the lengthy time in the cells prior to execution is cruel, unusual, and incompatible with civilized society.

"The ruling commutes the sentences of five women and one hundred and two men, including Robert Kennedy killer Sirhan Sirhan, as well as Charles Manson and his followers, who killed Sharon Tate and six others.

"A startling note for our listeners: with their new sentences, these killers will have a chance at parole as early as 1978."

I leaned against the porcelain, my eyes stinging from the trickling suds. The warmth of the shower turned to icy needles while the broadcast news festered in my mind and became a full-blown study of vengeance. "*Fuck!*" I lashed out.

A wet trail sprinkled below my robe as I stomped through the house searching for my mother. I found her in the family room, calmly sipping coffee, cradling Sharon's teddy bear in her lap.

On the table in front of her was the newspaper headline of the Supreme Court's vote, and next to that, an open Bible. I pointed at the paper. "Eight years for Sharon's life? If it's that fucking simple, I'll go and kill them all myself. I can't take another day of this crap!"

"Patti, don't talk like that," she warned.

"What is it you don't like, Mother, the word *fuck* or the fact that I want to kill them?"

"Darlin', your father just left here in a huff—with his gun, I might add. I really don't need two more killers in my life," she said, with quiet composure and a touch of sarcasm.

"How can you just sit there like nothing's wrong?"

"Let me tell you, it wasn't easy," she said, pointing toward the Bible. "Come sit down for a minute."

When Mom felt vulnerable and indecisive, she liked to open the Bible and let the pages fall randomly to a passage that would guide her. I looked to the scripture, opened to Proverbs 24:17. "REVENGE: When thy enemy shall fall, be not glad and in his ruin let not thy heart rejoice, lest the Lord see, and it displease Him and He turn away His wrath from him."

She placed the teddy bear in my arms. "Early this morning, I was in such a rage that I could have sold my soul in exchange for those creatures' deaths. My hatred was so strong that I'm positive it would have happened," she said, matter-of-factly. "But that choice would make me as dead inside as they are. Honey, let it go to God. Like the Bible says, He will take care of this."

"He didn't take care of Sharon, what makes you think He'll take care of this?"

"That's a question you'll have to answer yourself. I will tell you this, if you lose your faith in God, you've lost everything."

Doris

Right on schedule, the parole board granted Manson et al. a hearing before the end of 1978.

Between Manson, Atkins, Watson, and Krenwinkel's annual hearings, one of them surfaced in the news just about every three months.

Each time, the press called our house asking for commentary, and each time we remained silent, assuming that if we ignored the possibility of the killers' freedom, the situation would melt away.

Twelve years after the murders, the prospect of parole for the Manson Family not only endured, it made the headlines as a probability.

Before 1977, the state of California convicted prisoners with indeterminate sentences, or punishments that were largely subject to judicial discretion. Under that ruling, an inmate's sole prospect for re-entering society was in the hands of the Board of Prison Terms and Paroles. This sentence structure was based on the belief that a favorable parole review for release was persuasive incentive for prisoners to work aggressively toward rehabilitation.

During that time the California Board of Prison Terms ran as a seemingly unsupervised, governor-appointed group that used two indicators to decide an inmate's suitability for parole: prison modification and overpopulation. Those guidelines pressured the board into granting release dates to an average of three out of every ten eligible prisoners.

With the transition to determinate or fixed sentences, the crowded prisons threatened to burst. The California Department of Corrections, the CDC, viewed parole as their only release valve and pushed more prisoners with indeterminate sentences toward the parole board.

Attempting to alleviate their increasingly congested calendar, the board started granting an unusually large number of releases.

Deputy District Attorney Stephen Kay, who assisted Vince Bugliosi during the Manson trial, sensed the splintered freedom gates expanding. In an effort to repair the damage to truth and sentencing, Steve pioneered a program that encouraged prosecutors to oversee their convicts serving indeterminate life sentences by attending their parole hearings.

The initial prosecutors attending the lifer hearings found a common thread: the panel members did little, if any, research into the cases over which they presided. Disenchanted by that blasé attitude, the district attorneys persisted as public guardians and surrogates for the victims at the hearings to argue for continued incarceration of potentially violent inmates.

July 17, 1978

Steve made his debut appearance in the lifer program at Patricia Krenwinkel's hearing.

Prosaic to everyone involved except the prisoners wistfully musing what they'll do on their first day of liberty, parole considerations chronicle the offender's activities from their entry date into the CDC.

Cutting through the stale atmosphere, Kay's delayed entrance sparked everyone's curiosity, and in Krenwinkel's case, it provoked spiteful glances toward her past opponent. "Stephen Kay," he announced. "Deputy District Attorney, Los Angeles County. Sorry if I've held anyone up."

Commissioner Aquino nodded. "That's okay, we're just getting started." Then to Krenwinkel, "Prisoner will raise your hand and swear that the testimony you are about to give will be the truth and the whole truth?"

Krenwinkel's arm rose slightly. "I do," she said, then she pointed at Kay. "Is he allowed to be here?"

"The Community Release Board allows for his attendance," Aquino said.

Krenwinkel glowered. "He's just come to harass me."

"No, he's here to observe and then at the end of the hearing he has the option of making a short argument to contest or support your release."

"But he's—"

"Ms. Krenwinkel, the subject is closed." Aquino opened a two-inch file. "As I understand the statement of facts, on August 8, 1969, the prisoner, in the company of three crime partners, drove to Cielo Drive in Los Angeles and thereafter, as a result of their actions, five individuals were murdered." He removed his reading glasses. "Which of the victims did you participate in killing?"

"None. I never entered the house."

Aquino turned toward the sound of a guffaw. "Mr. Kay, you will not exercise those types of outbursts again."

Kay pursed his lips, and then nodded curtly. "Sorry."

The commissioner loosened his tie. "Ms. Krenwinkel, if you didn't enter the house, what were you doing during the period the victims were being murdered?"

"I was on lookout, near the gate."

Thoughtful for a moment, then Aquino asked, "The second night, August 10, did you stab Mr. LaBianca?"

"No," Krenwinkel said.

"What about Mrs. LaBianca?"

"I attempted to, but the knife I had, it really wouldn't stab her."

"How did Mrs. LaBianca end up getting stabbed?"

Krenwinkel shrugged. "Well, I know Tex went in. He was the only one that had a strong enough knife."

"Would you agree that Mr. Watson did a majority of the killing?" Aquino questioned.

"I couldn't tell you. By that time, I had taken over 350 LSD trips, so it's very difficult for me to remember the events surrounding that period."

"Did you know that you were going to murder these people?"

"No. I thought we were going to do a burglary."

Aquino leaned forward. "What about the second night? You knew that you were going to murder the LaBiancas, didn't you?"

"At the time I was frightened and did as I was told. Manson told me to go to the LaBiancas, so I did."

Forced to silence for the remainder of Krenwinkel's unchallenged adaptation of Linda Kasabian's role, the prosecutor's voice thundered in the cramped space when the board finally unleashed him. Kay had arrived at the California Institute for Women expecting to make a short statement. Instead, he took an hour to enumerate Krenwinkel's participation in the murders while the board members, with slackened jaws, listened to his unmitigated narration of the mayhem she'd enacted.

The parking lot was an inferno by the time Kay found his car beyond the prison's chain-link fence. He'd presented an assertive summation, influencing the system to keep Krenwinkel for at least another year. Nevertheless, he considered it a tenuous defeat within the Department of Corrections, where they incarcerate the average first-degree murderer for ten to twelve years. Time was running out.

He leaned against his car, waiting for the air conditioner to cool the interior. Perspiration beaded across his forehead. A breeze kicked up putrid air from the vast lands surrounding the prison. Out there, killing was a reasonable chain of events; however, when human beings traversed the line into the preying fields, could they ever be trusted again? He didn't have the answer, nor was it his objective to judge. His obligation as a prosecutor was to seek justice.

As he considered his options of keeping that integrity active, the same eerie feeling he had throughout the trials overcame him with the feathery hairs stirring on the back of his neck. He turned, sure that he'd find prying eyes upon him. Even so, Krenwinkel startled him. "Stay the fuck away from me," she snarled through the chain-link fence.

"And miss all this fun? That was quite a performance in there, Pat. Just like old times."

"You're the same as Bugliosi, looking to make a buck off this case. Well guess what, Mr. DA? I don't have to take your shit anymore."

"That's where you're wrong, Pat. I'm going to be here to make sure the only way you leave this dump is in a pine box."

She laughed. "Too bad it wasn't your house Manson picked. It's never too late for the little man."

"Is that a threat?"

"Why don't you go back to your seaside home in Palos Verdes," she spurred. "Correct me if I'm wrong, but it's right down the street from the Tate family, isn't it?"

He successfully hid the unease of her accuracy by moving confrontationally toward her. "Where I live shouldn't be a concern of yours."

She blew him a kiss. "Toodle-oo, Mr. Magoo."

Doris

I opened the front door. "Well?"

Steve gave a thumbs-up. "Krenwinkel was denied for a year."

"You don't seem too victorious."

"I've got to be honest," he said wearily, "I don't know how much longer the state will keep them in prison."

P.J. came up behind us. "I keep telling you, let them walk. I guarantee you they won't make it to the freeway alive."

I rolled my eyes. "My husband, the hit man."

"Mark my words," P.J. said, drifting away, "they'll never know what hit them."

"You want to come in for a drink?" I asked.

"No, my wife's probably holding dinner for me."

"All right. Thank you for going today." Then, as an afterthought, "Listen, if you ever need help, you know where to find me."

EARLY 1982 WAS the eve of the victims' rights movement in California. Crime hit an all-time peak and the trend began to shift with the public urging the legislature to implement rigid laws and tougher sentencing. Victims' organizations multiplied and gained strength in their united appeal for the same equal rights that the state afforded to the criminals.

The entire campaign wasn't even an inkling in my thoughts when Steve Kay walked into my hair salon. "Well, look what the cat dragged in. Come on in, honey. I've been waiting almost five years to get at that head of yours."

Steve allowed me to pull him into a chair. "I didn't even know you'd started the business."

"We just opened our third shop. I needed something to keep me busy, you know?" A curious thought hit me. "But you didn't come here for a haircut, did you?"

He smiled sheepishly. "Not really."

I felt the blood drain from my head. "Oh shit. Don't tell me they decided to release one of those guys."

"It's not that bad, but I do have a problem. Did you read the paper yesterday?"

"Darlin', I stopped reading the headlines years ago. Now quit beating around the bush."

"Leslie Van Houten has a parole hearing coming up, and she plans to present nine hundred signatures supporting her release. If I don't counter that support they might give her a parole date."

I got up and nervously cleaned the workstation. "There's nothing I can do about her; she wasn't convicted of Sharon's murder. Why don't you call the LaBianca family?"

"They're scared and don't want anything to do with it."

"We're all scared, Steve. Leslie Van Houten isn't my problem."

"Not on the surface," he said, "but if they let her out it will set a precedent for the rest of them to follow her through the door."

I dropped into the chair opposite him and looked around the sanctuary I'd built to buffer the troubles, as I called the curveballs God had seen fit to throw my way. On August 7, 1969, I was on top of the world—it was a long haul, but I was inches from having my family back together. Sharon had permanently moved to Los Angeles, I had a grandbaby on her way, and P.J. was weeks from retiring. I spent hours daydreaming plans for our first holiday season together in years.

On August 7, there were new beginnings just around the corner. P.J. was going to work on Roman's next film as a technical advisor. Jay had taught me his hair-cutting technique, and I was going to help with the distribution and promotion of Sebring grooming products.

Now, this place was all I had left to keep August 7 alive; and Steve had invaded it with the troubles. "I know I told you I'd help, but most days I can't even bring myself to believe that Sharon's gone, much less fight to keep those guys in jail."

"Leave the fighting to me. All I need from you is as many signatures as possible on a blanket petition against the entire Manson Family's release. It'll cover all of them so you won't have to do it again for the other hearings." He pulled a folded sheet of paper from his breast pocket. "I made this petition for you, but it doesn't have to be this formal, you can even hand-write them."

I reluctantly took the form. "I'm not very good at this kind of thing."

Cheryl, the stylist working next to us, reached over my shoulder. "Oh hell, give me that thing. I'd love to be the first one to give them a piece of my mind." After signing it, she pinched my cheek. "Cheer up. This is going to be a piece of cake. Everyone that walks through that door will want to sign this."

"KEEP YOUR NOSE out of it," P.J. said, before taking a swig of beer. "Let them go and I'll put an end to this whole goddamned farce."

"I've been thinking about it all day and I can't. They should have to sit in a four-by-four cell where they can think about what they did for eternity." I grabbed the beer from his hand and poured it down the drain. "And, if you stopped numbing your mind, you'd see that."

He reached in the fridge for another. "You're dead wrong if you think this will stop after you put those petitions in the shops. Sure as I'm opening this beer, the newspapers will catch wind of this and blow the whole thing sky-high. Goddamned reporters. I don't want anything to do with those sons of bitches."

"P.J., you're becoming a spiteful drunk, and I don't like it. All you do is pace around and grumble about life. You don't even see your friends anymore."

"I don't need friends. People are no damned good, and if you don't want to see this my way, I don't need you either," he grumbled.

"That's the beer talking, and if you keep it up, I'm going to let you die a lonely old man." On my way out, I flipped off the light. "There, that suits you much better."

"Oh, Mother," he laughed bitterly, "I'm already a lonely old man."

10

REVENGE OR JUSTICE?

Susan Atkins and Tex Watson have married, they have conjugal visits, they've each co-authored a book and say that they're very happy. When you stop to think of what they did, it says something about justice in America. If we define justice as giving a person his or her due, be it praise for a good deed or punishment for a crime, justice has been frustrated in this case. These killers have beaten the rap.

—VINCENT BUGLIOSI

Patti   1982

I walked out the front door and into the wall of Ohio's summer humidity. Even the newspaper felt sticky just minutes after its release from the deliveryman's car. A clap of thunder woke the baby. I looked toward the blackening cloudscape. Great. Another day in paradise. I dashed for cover, ahead of the deluge about to pour for the sixth day in a row. God, I was homesick.

The storm rumbled over the house while I rocked my crying eight-month-old daughter, Brie. "I'm with you on this one, baby," I nuzzled against her ear. "One more week and we can go home."

I stretched out on the couch with the paper, only to have a head-lined photo jolt me upright. KEEP KILLERS JAILED, SAYS SLAIN ACTRESS'S

MOTHER. Captioned below the picture: "1969, Sharon Tate's mother, Doris, and youngest sister, Patricia, mourn the day of the funeral." Hot coffee splashed out of my mug and onto Brie. She cried, I blotted, and my day hit rock bottom, all by 6:30.

For almost three years, my marriage had entitled me to a new name, a chance to start over, and a means to abandon the depressive persona of Patti Tate.

I had effectively become an outsider. I was now a traveling young mother, married to a pro–ball player whose name was the only one appearing in the papers—under the sports section. We were just a couple who mundanely settled into various suburbs with each trade to another city's team, socializing with a renewed circle of friends who didn't have a clue about the wife's sordid family history. At least that's the way her husband designed things to be.

Long ago, I traded in the drugs for a calming glass of wine, or usually a beer—I am my father's daughter, after all. These days, my childhood memories, good and bad, were well secured in a cavernous cellar that I seldom unlocked. On the occasions when I did work up the courage to open the cellar door, I stayed for only seconds, and I never dared pass the threshold to explore the gloomy catacombs, frightened of finding the bogeyman himself lurking, poised to attack with the ammunition of my childhood days.

God knows I love Sharon. But memories of her life brought sadness beyond repair; visions of her death, fear beyond comprehension. And so for emotional survival I'd packed only one thing from the past that I couldn't seem to let go of, my high school boyfriend's parting observation as our six-year relationship ended in 1978: "I don't think you'll ever allow yourself to be happy, and I can't live my life that way."

He may have been right; however, a coo from my daughter, or the feel of her tiny fingers frailly holding mine, had since brought me joy I couldn't have dreamed possible.

I cradled Brie on the right, as I held the phone on the left. "Mother, why are you doing this to us?"

"Doing what to you?"

"Stirring up this mess again. You turned in over ten thousand petitions, now for God's sake, leave it alone."

"You're beginning to sound like your father, and I'll tell you the same thing I told him. It's not a mess, it's justice for your sister."

"Justice? What are you talking about? All you're doing is making Manson newsworthy again." I read the article for the tenth time. "And what does this mean, 'Mrs. Tate has joined forces with Parents of Murdered Children as a victims' advocate?' "

"It's nothing. The group heard about the petition drive, and asked me to give a speech at their next luncheon."

"About what?"

"Surviving grief."

I laughed. "Who are you to lecture about survival? What are you going to do, compare shrinks? Lithium doses?"

"Why are you getting ugly, Patti?"

"I don't want my children growing up with this stain."

"I'm sorry you think of your family that way," she said, and then hung up on me.

Doris

My hand still rested on the phone; next to it was a pamphlet about Parents of Murdered Children. Beneath the abbreviated title POMC, their motto read, GIVE SORROW WORDS.

P.J. came around the corner. "I take it Patti saw the AP Wire?"

I nodded absently, turning the page to their timeline of the grieving process: *The first stage is numbness, a defense against being overwhelmed by trauma.*

"I don't know why you refused to warn her about it," he pestered. "I could have told you she'd blow a gasket."

"She called us a stain." *Next comes the most frightening and painful stage, called disorganization. Parents are covictims, and many of them get worse when the legal process is finished. Now, they begin to pine for their child in earnest. They realize he's never coming back. They have to recon-*

struct their whole belief system because the assumptions about the decency of humanity, the security of social order, and justice are all shattered.

"She's just rebelling, like the rest of us." He tapped the booklet. "We'd all be better served if you'd remember that you have children that are alive."

I swiped his hand away. "How dare you? I held this family together through days that I barely had the will to breathe. For thirteen years, I've taken care of everyone but me. I've listened to psychiatrists, priests, friends, and family tell me, 'Move on. You've got to get past all this. Start fresh.' I listened and I tried until I thought I was losing my mind because I couldn't forget." I waved the POMC leaflet under his nose. "This group understands me. It vindicates my emotions—*I'm a normal grieving parent.* What in the hell is wrong with sharing that experience?"

"Because it's none of anyone's goddamned business what goes on in this family."

"You're being ridiculous. I'm not airing our dirty laundry. I'm getting help." I stood to face him. "This is my journey, P.J. Come along or not. Either way, do it quietly."

Patti

Eight days of silence from my mother was all I could take. In front of my parents' house, I unhooked Brie from the car seat. "Not exactly the way I imagined our homecoming. Wish me luck."

An arm reached around, tickling my belly. Then Mom leaned in closer until we were cheek to cheek. "Breezy, honey, tell your mama she doesn't need luck, just some lovin'."

I turned into her embrace with a renewed appreciation for being a needy child instead of a responsible parent. I snuggled against her chest. "You smell like a bakery."

"I've got cookies inside," she cooed to Brie, "and if we hurry, we can get some before Papa eats them all."

"Too late," Dad said with a mouth full. "Where's that grandbaby of mine? The swimming pool's calling our names."

Like a million other California homes, my parents' backyard held a swimming pool, adorned by orange trees and night-blooming jasmine, all concealed behind a high fence of privacy. It was strange watching my father and daughter playing together. "He's good with her."

Mom handed me a glass of iced tea. "He was good with you, too, only you were too young to know it. Try not to resent him so much. He did the best he could under the circumstances."

"I love him; I just wish he wasn't so detached."

"Fleeting as they are, he has his gentle moments. You'll never catch him saying it, but he loves you with all his heart. So you all better start figuring out how to talk to one another because he may be all you have someday."

"You wouldn't dare leave me alone with him," I joked.

"Keep talking the way you did last week, and I might," she smoothly coursed us into our shallowly buried argument.

"I'm sorry for the cracks I made, but I haven't slept through the night since I saw that article. The whole thing scares the shit out of me. Every time you make the news, it's like a billboard advertisement to come kill the rest of us." I rubbed my tired eyes and pressed the frosted glass against an oncoming headache. "When all this started you told me to let God take care of it, yet here you are taking revenge."

She took my hand and pulled me up. "Come with me."

Inside the cabana, she pointed at the wall. "What do you see?"

I looked up and then averted my eyes. "A poster of Sharon."

"I know you don't like to think about these things, but look deeper."

I needn't look at it again, for the answer freshly lay on the tip of my tongue. The larger-than-life portrait was a long-standing friend of my youth that had quietly listened to countless confessions in Sharon's absence. Try as I may to erase painful memories, the dark honeycombs of my mind preserved every intricate detail of this photo that revealed my sister's essence.

When I grudgingly glanced up, tears brimmed over to rest on my cheeks. "Her eyes twinkle with the faith of her dreams."

"Yes," Mom whispered. "And when the ones that stole that from her go to meet the Lord, I'm sure He'll have something special for them. In the meanwhile, we have to do our work here." Her lips snubbed my tears. "It's not revenge, darlin', it's the price they must pay within the laws of man."

"All right. But here's the deal, not another word or picture of me or my family to the press."

"Fair enough," she said. "But tit for tat, in two hours I'm due to speak at that POMC luncheon and I could use the crutch of a friendly face."

It sounded like a miserable afternoon, but she caught me in a weak moment.

Dessert remnants and coffee littered the banquet tables. "Guess we missed lunch," I said, folding my arms protectively over my chest, half-expecting to find brain-sucking aliens within a group that gathered to talk about murder.

"Mrs. Tate?" One of them approached with blond hair, endless legs, and a nametag that marked her president of POMC. "I'm Connie. I was afraid you weren't going to make it."

"I'm sorry. I had car trouble. This is my driver, Gayle."

Speechless from her easy lie, I cleared my throat and smiled.

"There are two open chairs in the back. Lauren is speaking now, then you're next." The woman handed us a program before she flitted away as quickly as she'd come.

"Your driver?" I whispered out the side of my mouth.

"Shhhh," she nodded toward the petite woman whose head barely cleared the podium.

I scanned the leaflet. Lauren, single parent, whose only child, Holly, was beaten to death in 1980.

I looked toward the front of the room, where the microphone hardly amplified the voice of the once-doting mother. "Holly had just left home after dinner one evening. She ran out very happily on her way to meet a friend, and that was the last time I ever saw her. Through the long night I tried to convince myself that nothing was wrong, until a call came through from a detective who said, 'We think we have her body.' "

Mom wrapped her arm around my waist, and I slid closer into the shield of her girth while I reflected on a parallel diary. I looked around at a group that proved murder was unbiased in its pursuit. Deterred not by age, race, or social order, it hunted rather like a child: eeny, meeny, miny, moe. Watching these parents' haunted eyes, I knew what I'd do if it happened to my child, and it wouldn't be sitting in one of these meetings, nor would I ever attend a parole hearing—except for my own.

Lauren dabbed at her nose. "Her face was destroyed. They were unable to tell what color her eyes were. There were no teeth; her brains were out. . . . I desperately wanted to let go and go with her. I haven't had the guts. I really don't enjoy wallowing in self-pity. I'd been a mom for a long time. I still wake up in the middle of the night screaming, convinced for the moment that I have witnessed my daughter's murder. There's a terrible feeling of failure if someone kills your kid, because you didn't protect her. Nothing is easy anymore, but everyone here has helped me to bear the burden. Thank you."

Connie picked up her lead. "Bob and Charlotte Hullinger formed POMC following their daughter's murder in 1978. After their ordeal, they courageously set out to help other parents in need of emotional support and practical advice. Today, I'm proud to announce another courageous parent and first-time visitor, Doris Tate."

Mom didn't take the cue, in fact her arm anchored tighter around me. I nudged her with my elbow. "You'd better get up there before people start wondering why you're hugging your driver."

"What did you say this morning about being scared shitless?"

It was obvious only to me how nervous she was; especially noticeable when she used a ploy that I hadn't seen since my teenage couch lectures, as I used to call them. Then and now, she painstakingly rubbed her glasses clean with a tissue while she organized her thoughts. Scrutinizing her work, she put the lenses up to the light before placing them on her nose. "There, that's better," she smiled at her audience. "My daughter Sharon Tate was murdered by Charles Manson and his gang in 1969. It's taken me thirteen years to speak those words. God and the Blessed Mother give us the gift of denial, and I fluctuated between denial and panic for all these years. What Sharon went through the night she was murdered will never leave me. But now I have to be strong, because her killers have become eligible for parole."

She searched the diverse faces until she found Lauren's. "Your feelings are the same as all our feelings. We've all met the same challenges both personally and judicially. The dilemma is, what do we do?"

Mom shrewdly gazed at me. "The two questions I find myself asking lately are, when they're paroled, does my daughter come back to me? Does the punishment fit the crime? No. Your case, my case, it's all the same case—we've lost a child. And we've got a lot of work to do here to ensure that their justice prevails. Thank you."

A nurturer at heart, Mom milled through the group with a sympathetic ear to anyone who approached. I lagged far enough behind her to remain anonymous, and unfortunately, close enough to hear the shared stories. "At the morgue, I stepped off the elevator and walked right into my son's naked body lying on a gurney in the middle of the corridor. . . ."

"There I was, next to a chute, and like a side of beef, my daughter slid down in a plastic bag so I could identify her—although there wasn't much to go by. No one warned me that her face had been blown off. But there was a tattoo. . . ."

"It was so cold the night Nancy was murdered, and the boy that killed her left her with only one sock on. It rips at me to think of her dying all alone in the cold darkness. The police found a blanket five feet from her; if that boy had only thrown it over her, I don't think I'd hate him so much. . . ."

"I stood up in that courtroom and asked, 'What about Ginny's rights?' And do you know what that judge told me? 'Mrs. Thomas, Ginny lost all her rights the day she was murdered. . . .' "

"You're not really her driver, are you?"

I turned cautiously to see who was bold enough to address me.

"That's okay; you don't have to answer. I'm John Mancino—probably the only one here who hasn't been victimized."

"Listen, I'm really not interested in talking to a reporter—"

"No worries." He smiled. "I'm with a group called Citizens for Truth. We've done a lot of work with keeping high-profile criminals behind bars. Mrs. Tate looks pretty busy, and I have to leave. Would you give this to her?" He handed me an oversized envelope. "My number's in there if she's interested. And don't worry, your secret's safe with me."

FIRST THE MEETING and now gridlock traffic. I inched the car forward through the rippling heat that almost glowed above the freeway and gave me a craving for an ice-cold beer. Mom picked through the envelope Mancino gave me. "Who's the Onion Field Killer?"

"I don't know."

"This group raised twenty thousand signatures against his release. Same thing with Sirhan Sirhan. What was this guy Mancino like?"

"Didn't talk long enough to find out." My patience thinned.

"It says here that his group is looking to 'find solutions to our criminal justice system by electing to office officials committed to challenging the government into showing more compassion for the victims.' Listen to this, there's a victims' bill of rights—"

I put my hand over the page. "Mother, I can't listen to another word. I just spent two hours eavesdropping on stories about girls having their heads bashed in and mutilated bodies dumped in rock quarries. I'm so depressed I want to puke."

Moments passed in air congested with everything but conversation. I turned on the stereo. She turned it off. "You know, all these killers, they get to speak up and tell everyone how badly they want out of jail. Or how they've earned a degree. Or how they're a better person now. Their victims, including your sister, will never have that second chance. They'll never be able to say how much it hurt to die. We're their only outlet and damned if I'm not going to be Sharon's."

"I don't think she would want your life to revolve around her murder."

"You may be right, but she'd want to help others, and I can do that because her case is so famous."

"God, you're stubborn."

"And don't you forget it, my dear," she said, patting my knee before flipping the radio back on.

Doris

Later that night, I crawled into bed with John Mancino's packet. I shuffled through the papers until I found what had earlier piqued my curiosity: The Death Penalty Manual.

The California Public Defenders Association promoted the ten-page article as a guide filled with tidbits on achieving a victorious trial outcome.

Delay: A prosecutor usually wants to try a case when emotion is high. Time is a healing force. However, if you don't think you can delay it to a better disposition, then decide when you think the best possible time is to try the case; perhaps December, around Christmastime.

Engage in Paper War: Motions are trouble to respond to and prepare for. File a lot of motions in every case. You can even make motions challenging the configuration of the courtroom.

Rehabilitate your client: You need to be prepared to present all the good that can be said about your client and all the bad that can be said about the victim. District attorneys recognize that some victims need killing and if you can show that the victim in your case needed killing, even a little bit, this could be a real help.

I slammed the guide shut. With a red marker I scribbled across the title, *This must change or else Sharon will have died in vain.*

Sleep didn't come easily. I tossed and turned with the ruined lives I encountered that day still on my mind. Doubt crept through the strength I'd felt earlier until I eased away from the flowing thought traffic with the Serenity Prayer: *God, grant me the serenity to accept the things I cannot change, courage to change the things I can, and wisdom to know the difference. Let me live one day at a time, enjoying one moment at a time, accepting hardship as a pathway to peace . . .* I drifted off, thinking, then I just might make it through tomorrow.

LIGHT BLASTED THROUGH the back door of the limousine. "Mrs. Tate, I'm Carla, the stage manager for *Talk of the Town*."

"Hello." I reached out to the girl, who must have still been in college. "Honey, give me a hand, these things are hell to get out of; I don't know why they sent such a big car for one person."

"We send them for all our guests."

I shielded my eyes from the cutting glare of the San Francisco Bay that spread beyond the gray cinder-block building that housed the CBS affiliate.

San Quentin was a stone's throw away, and I wondered what Manson was up to behind those walls. I'd read in the tabloids that he said Sharon's ghost haunted him. The thought brought on a smile. *Give him hell, darlin'.*

The halls inside were equally drab-gray, save the posters of the local news personalities placed haphazardly along the walls. I trailed the quick-paced young woman, who seemed to be talking to no one in particular. "Pardon?" I asked.

She pointed to the headset of her walkie-talkie. "I was talking to the producer. You've got quite a following in there. Who are they?"

"Well, some are from Parents of Murdered Children and some from Citizens for Truth. How many showed up?"

"About twenty." Carla pushed at a thickly padded door that opened to a soundstage. "They're going to try to get in two shows this morning, so I'm taking you right to the set."

At the threshold, we bumped into a man lighting a cigarette on his way out. In defiance of the blue suit he wore, his ponytailed hair fell below his shoulders, forming an arrow that guided the eye to his biker boots. "Morning, Mrs. Tate," he said in passing.

I looked over my shoulder. "Who was that?"

Carla shrugged. "Beats me. Let me get this mic on you and we'll be all set."

"Mrs. Tate?"

I looked around, trying to follow the echo. The makeup woman powdering down the shine on my face paused her stroke to point up.

"I'm up here in the booth," the voice called out. "We're going to get started in a minute. Can you give me a quick sound check?"

"Hello. Can you hear me?"

"That's great. Listen for the music cue in sixty seconds, and then Catherine will come on the stage for the introductions. Try not to be nervous."

"You all are moving so fast I don't have time to get scared."

The tinny theme music played, "Welcome to *Talk of the Town!*" The host stepped onto the stage. "Our guest today is Doris Tate, who's here to speak out against the parole of the Charles Manson Family, serving life sentences for the murder of her daughter, Sharon Tate. Mrs. Tate, why are you against parole?"

"For a couple of reasons. First, public safety. Second, it's not a deterrent. In other words, which would make you think twice about murdering someone, the consequence of a seven-year sentence or a life sentence without the possibility of parole?"

The host stepped into the spectator's arena. "Before I turn the microphone over to our audience, tell us why you've started this campaign."

"Because I cannot depend on our elected officials to see that justice is served and society is protected. Our state paroles killers all the time; leaving these guys free to kill again, and our recidivism rate proves that they do just that."

A woman who didn't share my opinion took the microphone. "Don't you believe that people can change and be rehabilitated?"

"Perhaps there is room for rehabilitation in a sudden or impulsive crime of passion or anger, but not for the savage way Sharon and the others died. Susan Atkins and Charles Watson stood over my pregnant daughter and stabbed her sixteen times while she begged for her baby's life. I don't believe there's any hope of rehabilitation in creatures like that."

Some agreed and some didn't. Microphone or not, both sides squabbled their opinion in the free-for-all schemed show. An assertive baritone settled above the others; so much so, that it hushed the soundstage. "Mrs. Tate, maybe if you were better informed and knew all the facts you might have more compassion and understanding in the need to parole at least one of the Manson Family members."

I scanned the crowd until I found the antagonist. My heart fluttered a warning. Released from its confining ponytail, his hair draped over his shoulders. His hands clasped neatly over the blue suit jacket across his lap. He sat rigidly erect with a stare that seemed calm, yet his sweat stains said otherwise. He'd unwittingly chosen a seat next to my support group and three chairs from John Mancino. "How dare you speak to Mrs. Tate like that," John said, and then with a curious undertone, "You're one of them."

"Yes, I was associated with them."

The flutter from my heart turned into an all-out pounding that sent crashing waves through my veins; yet I was chilled as if the blood had entirely stopped pumping.

The frozen disquiet held for only a second before melting into hostility that rocked the bleachers. The two women sitting on either side of the man whose nametag identified him as Doc, shied away on his rise from the bench. "I have no anger toward you, Mrs. Tate," he managed above the protestors. "And I have no objection to your crusade; you're one of the few that has a right to speak out. The Manson Family only wanted to live in an isolated, peaceful environment, but the police were constantly harassing us, and beating us for no good reason."

Encouraged by my allies, I moved to the edge of the seat. "Peaceful or not, they will have to pay for their crime."

"But Leslie Van Houten didn't kill anyone. Rosemary LaBianca was already dead when she stabbed her."

"Now how do you know that?" I retorted. "Was the coroner standing there, counting whether she died on stab wound number twelve or forty-one?"

Doc took an aggressive step toward the stage, expressing himself in an equally set manner. "Well, I don't feel that justice is being served in her case. You need to let her out so she can put her life back together."

Fifty-eight-year-old grandmother or not, given the chance, I would have walloped this man a good one. "When will *Sharon* be able to put her life back together? When will *I* be able to put my life back together?"

Two security guards watched closely, ready to divert any physicality. When Doc took another step toward the stage, they grabbed his arms and hauled him back. "Keeping these people in jail won't bring her back," Doc ranted over his shoulder.

"Yeah, but as long as they're in jail they won't kill another who can't be brought back!" I said, watching the guards escort Doc off the stage.

Outwardly, I kept up the rock-steady bluff. Inside, the adrenaline subsided, and I trembled at my naïve oversight that Manson followers had been tracking my actions.

Carla stepped onstage. "Mrs. Tate, are you okay? Can I get you anything?"

"Not unless you've got a cigarette and a Tom Collins."

THE LIMO COASTED across the Oakland Bridge, headed for the airport. I smashed an umpteenth cigarette into the ashtray. "Dammit, John, Patti was right. I'm leading these assholes right to us. It's over. I'm not doing another show. It's not worth it."

Mancino was getting to know me pretty well and let me stew a few minutes longer. He toyed with the pack of cigarettes between us. "Want another?"

"Please." I took the lit Tareyton from him, but studied it instead of smoking it. "I had my first one of these at Jay's house. We went there after Sharon did this local interview. I was so mad at her that night because she'd worn this netted getup, without a stitch of undergarments. Mind you, there was little left to the imagination when she went on the television. So, anyway, there I was at Jay's, lecturing her about how her fame made her a representative of the entire family—and that outfit was not how I wanted people to think we acted." I twirled the cigarette. "She hands me one of these things, except it was the small European kind, and says, 'Here, this will calm you down.' And you know what? She was right—usually was when I gave her half a chance.

"Dammit, I shouldn't have been so hard on her. I'd give anything to take those words back." An exhaling cloud trickled out the cracked window as I wrestled with guilt. "What I really wanted to do at the time was wring Roman's neck. I blamed that man for everything because once they were together, she seemed distant from the rest of us—didn't even get invited to the damned wedding—and it all started with that God-awful *Playboy* spread Roman had her do. Oh, if you could have seen P.J.'s face," I laughed. "But I've lived and learned in my old age, and the thing is, we're all responsible for our actions; good, bad, or indifferent, we own them. Now, take Watson and Atkins. They love to blame Manson, you know. But they did it. *They did it.*"

Mancino perfectly timed his pitch. "So, are you still going to walk away and let them win?"

"Hell, no."

"Good. The *National Enquirer* called."

"Why? Did they find Sharon's baby alive and well in the Everglades?"

"Laugh it up, but they're offering to help with your petition drive."

"I'm not doing that trash."

"It may be trash, but they've got more than one million subscribers and millions of readers; if only five percent signed a petition coupon, we'd get over a hundred thousand."

"Well that certainly makes it interesting."

"Watson's hearing is only a month away. Imagine going for your impact statement with a truckload of petitions."

"I'll deliver them, but I'm not going to the hearing. Nope. It'd be torture to look at any one of those killers, and believe me, I can be more effective if I keep my sanity."

ON THE ADVICE of the "Death Penalty Manual," just after Christmas, on December 28, 1982, *National Enquirer* fans read the petition drive and their response was overwhelming.

A full moon reigned in the silky blue sky when Mancino and I arrived at the California Men's Colony, not in a truck, but a van filled with 249,000 petitions opposing Tex Watson's release. We pulled into a space close to where the press pool set up near the entrance to the prison's administration building.

I wiped a frayed tissue across my raw nose. "For two hundred miles I've been asking why? Why did Sharon have to die the way she did? Why didn't I watch over her more carefully? Why couldn't I be one of the lucky parents?"

John scarcely broke a whisper. "I don't know."

"I thought this was going to be a great day, but I feel like shit."

"Listen, you'll get through this, just like you've gotten through everything else. Besides," he motioned toward the reporters nearing the van, "it's showtime."

Before my feet hit the pavement, a journalist reached me with an outstretched tape recorder. "Mrs. Tate, why are you doing this?"

Buying time to work up the pleasant conviction I had learned to display for the media, I smoothed my skirt, and then looked in the van for my purse, which was easier found than I'd admit.

I could have stalled another week. It wouldn't have made a difference; subdued was the best I had to offer. "I think it's horrible that I have to solicit names to keep Tex Watson in jail. Let me ask you a question. Why is it that a man who was sentenced to die is waiting right inside that building for his chance at freedom? Can you tell me that?"

"Will you attend the hearing?" a different reporter asked.

"No. I would feel like I was doing this for revenge and that's not what this is about."

"Mrs. Tate, I've heard you say that thirteen years isn't enough prison time for Watson. When will it be enough time?"

I pulled a picture of Sharon from my purse; holding it high enough for everyone to see. A knot swelled in my throat. "This is all I have left of Sharon; this and a headstone that I can talk to a couple of times a year. It will *never* be enough time."

THE PRISON GUARD closed the lid on the last of the thirty boxes he'd searched. "Sorry for the delay, Mrs. Tate, but it's policy; nothing comes through those doors without being searched."

"That's okay. Tell me, is this the room where they'll conduct the hearing?"

"Yes, Ma'am. See the camera there? That's connected to the press pool outside. I'm going to get Mr. Carter; be back in a jiff."

John waited for the door to completely close behind the guard. "How are you holding up?"

"Like you said, I'll get through it. Luckily, I won't be in this shoebox with that bastard."

"We should at least stay and watch the hearing from the press pool."

"The thought of Watson roaming around these halls has me jumpy as a frog. Nope. This is a little too close for comfort."

"Doris, how can you expect to effectively help others or lobby against parole if you don't know what happens during the process?"

A key clanged into the door lock. A handsome man who looked like he'd be more at home on the golf course than the prison stepped inside. "Mrs. Tate? Board Superintendent Bob Carter. This is for you. It's a receipt of sorts that says the petitions are now registered and a part of inmate Watson's permanent file. Will you be staying for the hearing or shall we notify you of our decision by mail?"

"I guess we'll watch from outside at the press pool," I said, leering in John's direction.

"Good enough. I'm sure our paths will cross again," he said, and turned to leave.

"Mr. Carter? Don't forget, after your decision today you're going to have to look in the mirror and know you did the right thing."

"Mrs. Tate, I assure you, I do that every time."

SUSPICIOUS AS I was that Tex Watson peered at me from behind one of the barred windows, my eyes shifted from the television monitor to the surrounding prison buildings, unsure which frame his face would first appear in. I pulled my chilled arms in tighter and stretched my neck, trying to quiet the ache thriving between my shoulders.

The cocktail party atmosphere of the others, mostly reporters, who waited for the hearing to begin, gave me cause for envy. In a couple of hours, they'd all be racing toward their next story, their time here forgotten, while I remained perpetually imprisoned to await the next hearing.

Of course, I held all the killers responsible for Sharon's murder, but over the years, I'd built a particular loathing for Watson. Where I come from, which happens to be the same place Watson is from, men protect women at all costs. It was beyond comprehension that a man could kill a pregnant woman, let alone with the zeal Watson practiced on his victims.

Movement within the monitor screen caught my attention. Two men, besides Carter, spread their folding chairs against one side of the table. Stephen Kay took a seat at the short end. Aside from the two spots reserved for Watson and his attorney, three media representatives, a cameraman, and two guards packed the windowless room. The six microphones spread out on the table were obviously not for amplification purposes. Carter tapped his, and then slipped an audiotape into a decrepit machine. "Today is Thursday, January 13, 1983. We are at the CMC East, San Luis Obispo, to conduct a subsequent parole consideration hearing for life prisoner Charles Watson. CDC number is B-37999, case number A-253156."

The door opened again, and Watson sauntered in behind defense lawyer Olpin with a gait just shy of cocky. The former Texan kept his hair modishly coiffed right at the point where his pale blue shirt collar met with the like-colored sweater. A bushy though immaculately trimmed mustache covered his thin upper lip. The camera zoomed in to capture his steel-blue eyes. Before Watson stabbed Sharon he'd said, "I'm the devil, here to do the devil's business." I gauged his icy eyes, sure that he was correct.

"Mr. Watson, we usually break down the discussion into four

parts," Carter began the hearing. "I will lead the discussion concerning the commitment offense and your prior social and criminal history. Mr. Tong, on my left, will lead the discussion in post-conviction factors. And Mr. Lopez, on my right, will lead the discussion for your parole plans. Although we will take turns leading the discussion, all of us can ask you a question in each of these areas, including Mr. Kay, representing the people of Los Angeles County.

"In prior hearings, we've developed a statement of fact that describes, in a very succinct way, what is a very complicated description of the occurrences that led to the deaths of seven victims. I am recommending that we incorporate that reference instead of rehashing through it. Are there any objections?"

"No objection," came Mr. Olpin's response.

"I'm agreeable to that," Kay said; "however, I would like to make some corrections. On page five, talking about Mr. LaBianca, it says, 'A long cord was wrapped around his neck and a carving knife was in his abdomen.' That isn't correct. A carving fork was in his abdomen, and then we should add that he had a carving knife that traversed through his neck.

"Then, in the last sentence, it says, 'Mrs. LaBianca suffered forty-one puncture wounds.' They were actually stab wounds. Mr. LaBianca was the only one that suffered any puncture wounds from the carving fork in his abdomen."

"Well, is there really a difference between puncture and stab?" Carter asked.

"I believe the victims would argue that the eight-inch blades used to murder them would constitute a stab wound as opposed to a puncture wound," Steve said. "Finally, on page four, where it says, 'Miss Tate was hung,' we need to add, 'while she was still alive.'"

Inside the boardroom, they reviewed clerical errors. Outside, I shifted the medal on the rosary one bead over and said a prayer. Instead of a communion with God, the blinds of my eyelids ignited a phantom's view of Sharon suspended. Hands bound at her back, she thrashes against the snare, her mouth gapes open to scream, yet she can't. This is not a new vision. I've seen it—and worse—a thousand times over. My grip tightened around the beads as I shook the scene away; even so, a shiver held tight. John looped his arm through mine. "You want to take a walk until they finish with this part?"

"Not yet. I'm going to take this one minute at a time and see if I can keep from cracking," I smiled weakly.

"Mr. Watson," Carter asked, "would you concur with those modifications?"

"Not exactly, no."

"In what areas would you disagree?"

"Well, I'd disagree with the fact that Sharon Tate was hung while I was present."

"Do you deny hanging her at that time?"

"Yes, I do. I haven't denied too much, in fact I've confessed to all of it, but I don't remember her being hung. And I have often thought about this, and I don't know how that happened, but I have no comment on that. I just don't remember that particular fact."

"Are there any other comments in this area?" Carter looked around the quiet table. "Okay. We're going to move on to the criminal's history, which is very short. Two arrests: one while in college for the burglary of some office equipment, and one prior arrest in California for being under the influence of drugs.

"Prisoner had what could be described as a sterling social history with a stable family life, lots of community and school involvement, and ample opportunity to turn out to be the average John Q. Public. Mr. Watson, do you have any insight on how you turned out to be something so grossly different?"

"Well, 1969 was a very revolutionary atmosphere. I was very rebellious to society. And I was seeking something that I had not found for satisfaction, for happiness, for contentment. So, in my search, I began to take drugs and hallucinogens. And more or less dropped out of society into the love generation, although it turned out to be very much hate.

"I was searching for acceptance and the answer in life. And

I thought I'd found it through LSD and Manson, along with the Family living at the ranch and all the love that I felt there. And by taking the drugs for over a year, I lost all sense of care and direction in life. I had completely forgotten about my background, my past, my upbringing. My parents no longer existed in my thinking. I became very susceptible to the Family and the teachings of Mr. Manson. And gave myself completely to him, to what the Family was doing. To a point of where I would do anything without care for myself, my own personal safety, my own life. I had very little feelings for anybody.

"There was no right or wrong with Manson, there was no time and space. The people that I killed didn't even seem real to me. At the time, I saw them as blobs. And I thought that in killing them, I was only killing myself. Right after the crimes, I thought the end of the world was coming because of a black and white race war."

Carter shook his head in disagreement. "I think it's a common phase of life for kids to want to break away from their parents, but the extreme turn that you made is the difference between you and hundreds of thousands of other people. All right, let's move into the area of postconviction factors. Mr. Tong?"

"Let's see," Tong scanned a sheet, "you have recently taken and completed a psychology course at Cal Poly?"

"Yes, with an A grade."

"Okay. For two and a half years, you held an assignment as a clerk in the psychiatric intervention unit and then as a student chaplain. On April 21 of 1981, you became an ordained minister with the Word of Faith Church in Bakersfield. What were the requirements to become a minister through that church?"

"It's a ministry training program that you begin as a congregational member, and then you have the availability of going into the ministry under the guidance of Reverend Stanley Maguire."

Tong pried a page from Watson's file. "This is a letter from Reverend Maguire that comments on your involvement with the church here. He writes, 'Mr. Watson is my associate pastor in charge of our worship department, student chaplain program, yokefellow group therapy, and works with the administration.' "

Watson nodded enthusiastically with each of the notations. "Yes, I believe that's an accurate interpretation."

"My final note for this section concerns the legally incorporated ministry you have, called the Abounding Love Ministries, Incorporated. At your last hearing, you had approximately fourteen hundred people that you were in contact with. Has that number grown?"

"At this present time, we're in contact with thirteen hundred prison chaplains throughout the United States and Canada. We also have twenty-five hundred to three thousand people on our general mailing list that we minister to."

"Any other comments for this part?" Tong asked.

"Well, I learned something this morning that concerns me a great deal," Steve Kay responded. "I don't know what's happening in the prison chapel here, but I was informed today that Mr. Watson and Bruce Davis are working hand in hand in the chapel. Now, Tex Watson and Bruce Davis were the two main leaders of the Family when Charles Manson wasn't there.

"You've got to remember that the Manson Family was a quasi-religious group to start with, so the fact that Mr. Watson and Mr. Davis are suddenly born-again Christians is not surprising. And now, to allow them to be together as leaders of this chapel, well, it's a situation I'm very much opposed to."

Watson's southern charm faded with the blink of his stony eyes. "Personally, I don't think this hearing has anything to do with Mr. Davis and my association, or our Christian experience. I think that Dr. Stanley Maguire is highly capable of making decisions that shouldn't concern Mr. Kay.

"Mr. Davis is a legitimate born-again Christian, and we have a very good relationship. Matter of fact, we live together on the honor unit. I don't have a lot of recollection to what degree religion was used in our fam—*the* Family, we never discuss it. So, what that has to do with this board hearing today, or Mr. Kay, I don't fully understand and, well, I wanted to make that point clear."

Watson's outburst gave pause to the three panel members, while Steve Kay sat back, seemingly satisfied that he'd pushed the right button. Carter spoke first. "That certainly does clear things up a bit. Would you like to go into future planning, Mr. Lopez?"

"Sure. But first, I'd like to ask Mr. Watson what his thoughts are on his progress toward a future outside this penitentiary."

"Right after I was convicted of that crime, Mr. Bugliosi said that the only thing that could help me was a new heart. Until I gave my life to Christ and began to go down the road of Christianity, and began to mature my Christian walk, I never felt any remorse for that crime. In 1975, when I gave my life to the Lord, I felt remorse for the first time—I received that new heart. Since then, I've begun to mature in my thinking and my psychological makeup. I know, without a doubt, that my life has gone through radical changes since I've given my life to Christ."

"Okay. Next I have a report from August," Lopez said, "that indicates you were married in September of 1979, and that marriage is still intact. That you have one child . . . who I believe is four weeks. . . ."

I clutched John's leg. "Dear God. Did I hear that man correctly?"

Mancino numbly shook his head.

"Impossible. I mean, he is in prison, right?"

"Must have happened during one of his conjugal visits."

"What in the hell is that?"

"California prisoners are allowed to have unsupervised, forty-eight-hour visits with their spouses—trailer sex."

A reporter edged toward us. "Mrs. Tate? Any comment about Watson being a parent?"

I stood to leave. "No." Then to Mancino, "Get me out of here before I vomit in front of all these nice people."

And at the van, that's exactly what happened. I sat back in the passenger seat. "Okay. I think I'm finished."

"Should I get you a Coke or something?"

"No, just drive. I want to get as far away from this place as possible. How dare Watson enjoy the very pleasure he stole from Sharon? It's a good thing I wasn't in that boardroom. I'll tell you, when I heard about that child, I would have broken every bone in his body. You know, I knew that Watson living in a cold, dark dungeon was pure fantasy, but I never imagined all this; no, sir, not in my wildest dreams."

At seventy miles per hour down the freeway, the citrus groves streamed across the window with hypnotic hues of orange, yellow, and green while I combated the analogy of revenge versus justice. When I finally had my finger on it, I patted John on the shoulder. "Get ready, because this is war—not revenge, mind you. My definition of revenge would be taking away something Watson loved. Nope, this is justice. And by that, I mean he's going to go through another radical life change, because I'm going to fight him with everything I've got. And I don't care how long it takes, but he's going to start living like a convicted murderer." I rested my head against the seat. "Conjugal visits my ass," I muttered.

11

THAT OLD BITCH

When I hear people speak about the death penalty, asking isn't it cruel and unusual? I have to say, absolutely not. You want to see cruel and unusual? Let me show you what they did to my sister and all the other victims. What they did was cruel. We think so much more about the killers than they ever, ever gave a thought about their victims.

—PATTI TATE

Doris

There was a young girl named Amy Sue Seitz whose story, though unrelated to my own, has tremendous bearing on mine just the same. I came upon news of it in 1978.

By the time little Amy Sue left this world, the police explained to her mother, it was a blessing.

Mrs. Seitz's ordeal began six days earlier and fifty miles north of Los Angeles, where the populace dropped to seventy-five thousand in Ventura County.

Without fail, the spring months in California set free the orange blossom scent of heaven to reward its residents. March 14, 1978, was just such a morning when a few minutes past eleven o'clock, Amy

Sue toddled out the front door of her aunt's home and disappeared without a trace.

Almost one hundred officers combed the streets in search of the missing two-year-old. So many helicopter pilots volunteered for the search that it looked like wartime as the fleet of aircraft hovered over the small suburb of Camarillo and its surrounding areas. Neighbors and their hunting dogs joined the canine units to search the weed-stricken fields that would have towered over the thirty-two-inch child.

It was a valiant search; nevertheless, fruitless. Before the first helicopter launched from the airport, Amy had left Ventura County.

On the vinyl backseat of the car, Amy didn't have the strength to wriggle free from the bounds around her hands and legs. Her underwear, wedged into her mouth, muzzled her cries. Wide eyes, no doubt moist with replenishing tears and fear about what was to come, she watched the operator of the vehicle propel them southbound on the Ventura Freeway.

An hour later, just off the Topanga Canyon exit, Theodore Francis Frank discarded his latest toy into a ravine. It was Amy's body.

The coroner had a difficult time comparing the remains to a portrait Amy's mother gave him for identification. In fact, it couldn't be done. The blond-haired, blue-eyed little girl who posed for the picture in no way represented what lay on his autopsy table. Black holes filled the space where her eyes once shined, and she'd been beaten so severely about the head that the surrounding skin only loosely bound her skull. Closer review proved that her attacker sodomized and raped her.

In the course of his career, the dead never sickened the examiner, but when he removed the torn, blood-drenched clothing, he felt a wave of nausea. Whoever molested the infant had peeled the skin away from her buttocks and hips. Both of her tiny crushed nipples, imprinted by vise-grip pliers, dangled from her chest. The missing teeth that the doctor had earlier noted, he later found in her stomach, along with the equivalent of three beers.

Tragically, the preceding notations weren't the cause of death. Amy survived it all, until Frank strangled her.

Before Amy encountered him, Theodore Frank had been arrested six times and spent fourteen years locked up over the course of his twenty-two-year vocation in child torture.

Due to his criminal history, his defense attorney managed to have his last conviction reduced to an MDSO—Mentally Disordered Sex Offender.

Defendants like Frank who fell into the MDSO program were immune to prison sentences. Instead, they went to a state mental hospital for eighteen to twenty-two months, after which the doctors deem them rehabilitated and sign their release papers.

California sent Frank to Atascadero State Hospital for a two-year evaluation.

Every minute a parent waves good-bye to their child telling them to be careful and don't talk to strangers. "Okay," the innocent yell over their shoulder as they run off. Yet few are considering the warning. The ones that do envision a wild-eyed, drooling monster, not gaily dressed clowns like child murderer John Wayne Gacy. Nor are they expecting middle-aged men like Theodore Frank, with peppery hair, beard, and mustache. Dressed immaculately in suits that he accented with thick-rimmed glasses, Frank appeared more the scholarly Dr. Jekyll than his alter ego of Mr. Hyde.

At Atascadero, Frank equally fooled the psychiatrists with claims of remorse and rehabilitation.

Behind the scenes, the men in his ward knew him as a braggart for sharing explicit tales.

In the privacy of his cell during his two-year stay, Frank relished his attacks by composing a journal detailing the sexual torments he committed against one hundred children.

The recidivism rate for an MDSO is 99 percent. Frank was no exception. Six weeks to the day of his release from Atascadero, he killed Amy.

While Frank eluded the detectives, a terrified neighbor held the key to unlocking the case. The police had tried to talk to her within hours of Amy's disappearance. At the time, the Mexican mother of two toddlers claimed ignorance and slammed the door on the officers.

Winter rains threatened to come early when the detectives decided they had nowhere left to go in their investigation but back to the beginning to reinterview the neighbors. In a repeat performance, the woman said, "No hablo Ingles," and closed the door on the inquiring policemen. But then, the door opened again. This time it was a man. "My wife has something to say. The day before that little girl disappeared, I think she saw the man you are looking for. He tried to take my daughter for ice cream."

Out of three hundred and fifty mug shots of convicted molesters, the woman identified Frank as the man with her daughter.

The Camarillo detectives caught up with Frank at the Los Angeles County Jail, where authorities held him on charges for assaulting two other children.

Boastful as ever, Frank confessed Amy's murder to a fellow inmate in the Los Angeles jail. "I've done this before, over and over. It's like, you get going, doing it, your adrenaline builds, the excitement is there. With this kid, I let things get out of hand, that's all."

The investigators obtained a meticulously itemized warrant for the evidence they were looking for in Frank's apartment. Number eight on the list specified: "Scrapbooks, newspapers, photos, tape recordings, or writings which could relate to the death of Amy Sue Seitz."

The detectives found incriminating evidence hidden throughout Frank's apartment. They cataloged pliers, newspaper articles of his crime, a gas receipt that placed Frank in Amy's neighborhood on the day of the abduction, and finally the journal from Atascadero.

The items seized during the search were just a sampling of the powerful evidence presented during Frank's monthlong trial.

On December 15, 1979, following the jury's guilty verdict, Judge Byron McMillan sentenced Frank to death. When he set the penalty, McMillan commented, "Mr. Frank, I wouldn't sweat it. Unfortunately you will probably die of old age—out on the streets, still molesting children."

After an inmate is sentenced to die in California, their case is automatically sent to an appellate court. McMillan based his prophecy on the infinitely slow process of appeals to the State Supreme Court that often take more than twenty years. In Frank's case, his death sentence was overturned within five years.

Based on the Fourth Amendment, Chief Justice Rose Elizabeth Bird led the State Supreme Court's decision in Frank's penalty reversal, stating that the police violated Frank's rights to privacy by seizing his personal journal without justification.

Bird drafted their official decision: "This warrant was too overbroad a description. Although the officer went through the motions of obtaining a warrant, his actions prove he used it as a license to get inside the defendant's house. Ignored the list of property specified. And had no more interest in the warrant than a theatergoer with his ticket, once he's been seated."

With the journal considered an illegally seized document, the high court ruled that Frank's jury had been improperly influenced by Frank's writings. In accordance with their decision, Bird said, "The Supreme Court has not gone far enough to protect the criminal's rights. If it was solely up to me, I would reverse Mr. Frank's conviction as well—the diary was unfairly used in the trial."

Doris   1982

That same day of Frank's reversal, the high court vetoed three additional cases, two death penalties, and one murder conviction.

I cut out the newspaper photo of Amy and put it in my billfold. It was an election year for the justices of the supreme court, and that little girl became my impetus to unseat those I believed to be exceedingly biased jurists.

Like Frank, I kept a journal of my own. By 1985 I'd reached into the jar of every victims' rights issue and advocacy group I could find, becoming a board member for Citizens for Truth, Justice for Homicide Victims, the California Justice Committee, Believe the Children, and took over the position of president at POMC.

I divided my days counseling victims, lobbying legislators, giving speeches, and appearing on talk shows and news programs. The few moments left open, I put into the hair salons.

If we lived on a ranch, I would have already put in an hour of work and three cups of coffee before hearing the rooster's warning of sunrise. During that time, I made the day's to-do list. With the morning headline of the supreme court's ruling on Theodore Frank's case, I made only one notation in the journal: "Concurrent sentencing is farcical—ten counts of child molestation with a sixty-year sentence equals two years. Who is Chief Justice Rose Bird?"

Born in 1936, Rose Elizabeth Bird was raised by her widowed mother in an underprivileged environment. In spite of the odds, Bird developed into a woman of groundbreaking achievements.

Upon completing her law degree, Bird was the first female clerk to serve the Nevada Supreme Court. In Santa Clara, California, she became the first female cabinet officer. And on March 26, 1977, she became the first woman appointed chief justice in the California Supreme Court. Bird was also the first and only chief justice to overturn every lower-court death sentence that came to her appeal.

Bold, articulate, and unwavering in her opposition to the death penalty, Bird had vetoed fifty-nine cases. "What looks like a technicality to some people is a right to others," she noted of her decisions. "My role is to do what's right under the Constitution. We are the guardian of rights, and we often have to tell people things they do not like to hear, and if that's politically unpopular, so be it."

Somewhere along the line, Rose Bird forgot about the victims' rights and decided that "We the People" encompassed a small group of jurists.

We may roost on opposite sides of the fence, but Bird and I were alike, each deep-seated in our convictions, each influencing our perceptions on the public.

A year earlier, I ran for the state assembly in the Fifty-first district, Palos Verdes. Touted as the "Law and Order Candidate" and criticized for my campaign's focus on victims' rights, I lost the race by thirty-five percent.

Following the defeat, I reflected that I didn't care about being a politician; rather I had pursued political power's ability to protect victims and potential victims from what I perceived as the leniency of the judicial system.

The experience left me with a valuable lesson—personal agendas don't belong in the political arena.

The supreme court's four overturned convictions happened on June 6, 1985. By the following Monday, embers sputtered wildly from the political fire of opposing forces, and with a gust, I fanned the flames.

Proponents of Rose Bird claimed the campaign to unseat her was a right-wing effort to politicize the court, reduce its independence, and subject its decisions more to the public's whim than to the law.

Bill Roberts, the leader of a statewide organization that campaigned to relieve Bird of her duties, noted, "The chief justice has blatantly and systematically exceeded her lawful authority, and has been imposing her own radical political philosophy on every resident in this state; using her judicial position to legislate rather than to interpret state law."

My reasoning was a bit more colloquial. Shit rolls downhill; if the supreme court won't uphold the lower court's decisions, there's nowhere left to go. Until we get the right players on the field, why enter the game?

It was an ambitious plan: 1934 was the last time the state of California unseated a member of its supreme court; a chief justice had never been ousted.

Election Day closed in. Together with Roberts and a multitude of others, I organized a press conference at the State Capitol. The first to address the gathered media, I picked up the microphone. "I'm here this morning to add my voice and energy to defeat Chief Justice Rose Elizabeth Bird at the polls next week. Her record of interpreting the law to favor the defense, and the rights of violent felons over the safety of the public is a matter of record since the time of her unfortunate appointment.

"Bird's comments that we, her opposition, consist of only right-wing extremists is simply fantasy. For as it has been said, we the victims are Democrats, Republicans, and Independents who recognize the need for change. Change that will again bring sanity to a system that has failed to control violent crime. A system that makes excuses for continually releasing violent felons. The defeat of Rose Bird is an essential start toward safeguarding society from the criminally violent."

The California debates aired nationally, where the issue narrowed to focus on the death penalty. Those resistant to California's gas chamber called it barbarically inhumane murder that is applied unfairly, arbitrarily, and discriminatorily to its victims. Those with that frame of mind also argue that it's biblically unethical, since it breaks one of the Ten Commandments: Thou shall not kill.

To paraphrase prosecutor Logan Green, I saw nothing to be gained by arguing the Bible or any other interpretable issue. Our state provides that murder is punishable by death and our judges and juries are there to enforce it. Nevertheless, since Rose Bird did the talk-show circuit seeking support at the polls, I followed suit to contradict her efforts.

Phil Donahue asked, "Mrs. Tate, why is the death penalty important?"

"Too often death sentences are modified, then shortened, and before we know it, these murderers are back on the streets to kill again. If our courts aren't willing to apply truth in sentencing, and if we don't have the means to keep these repeat offenders incarcerated, then we must carry out their sentence of death. We must have deterrents."

"Many would disagree with you; statistically, the death penalty has not been shown to be a deterrent to crime," Phil noted.

"Of course it hasn't, because we don't enforce the death penalty. Californians have sent three hundred and forty-six killers to death row in the last twenty years, but only one inmate has been executed.

In California, each week, one prisoner sentenced to life behind bars is released. Listen, Phil, we could debate the deterrent question to our graves, but one thing is certain, it *will* cut down on recidivism, because the guy that goes to the gas chamber, well, my dear, he's one less we have to worry about."

"Is the death penalty cruel and unusual punishment, Mrs. Tate?" Merv Griffin asked.

"Honey, I want you to walk a mile in my shoes and then ask me that. Wasn't it cruel and unusual when they killed their victim?"

In November 1986, the people of California voiced their insistence for impartial honor as they never had before. Chief Justice Bird was defeated at the polls along with two other pro-defense judges, Cruz Reynoso and Joseph Grodin.

Although there were celebration parties to attend all over Sacramento, I opted to watch Bird's concession speech.

I stood at the front of the somber crowd as Rose Bird took her position in front of the cluster of microphones broadcasting throughout the nation. Her blond hair fell to an elegantly draped scarf that adorned her perfectly tailored black suit. The former chief justice brushed a wisp of her fallen bangs; uncovering her sad, yet dignified, glances to her supporters.

For a moment, I thought Bird might break down. I almost felt sorry for her. Almost. Studying my long-term foe, I didn't relish the idea that I'd participated in the destruction of her career; however, there are always casualties of war, and she'd been enemy number one. With that thought, Bird and I locked eyes for the first time. We each knew the other, yet didn't at all.

As if in defiance, she focused on me until the completion of her first sentence. "I appreciate that some people within our state are impatient, impatient to see executions. But I say to those who voted for us today, that although my voice will go silent, yours will not. You still can fight to ensure that we retain this house of justice. I don't think anybody in this state will sit easy if in fact this becomes a court that ensures nothing but executions."

I ambled through the crowd contemplating Bird's final speech. I cringed at the idea of a chief justice who executed with the enthusiasm that Bird overturned convictions. It was not the solution. I glanced up at the cheerfully blue sky, my face reaching for some of it to rub off on me. Six months ago I'd joined Believe the Children, an organization for abused and molested children. These same children needed to be able to believe in grown-ups as well, grown-ups who will protect them from the Theodore Franks of this sometimes-nasty world. Today was a victory for Amy Sue Seitz. I blew a kiss toward the sky, hoping she was close enough to catch it.

In February 1987 Theodore Frank received his second penalty trial, sans the journal. The evidence beyond Frank's writings was so overwhelming that it took only four hours for the jury to deliberate a death verdict.

Later the foreman said, "The jurors were never close to voting for the only other choice before them: life without parole."

In September 2001 Theodore Frank died of a heart attack on death row.

Doris

"That old bitch needs to shut up and mind her own business. She doesn't know what she's talking about; *I didn't kill her daughter* and she knows it."

The clip of Charles Manson faded to black while the lights in the studio brightened. "Doris, does Manson evoke a reaction in you when you see him doing these types of interviews?" Geraldo Rivera asked.

"Uh-huh," my eyebrows rose to match a playful smile. "I want to wring the little—well I can't say it on TV, okay?"

"It's been a couple of years since we last talked," Rivera commented. "Have you changed?"

"Well, I have five years' more experience in doing this." I laughed and then turned to the audience. "Did you all hear Manson call me 'that old bitch'? That's his pet name for me; a name, I might add, that

I wear with pride, because it means that I'm getting under the little bugger's skin."

"And what about the other killers? Update us on that, if you will. They all look so cleaned up."

"Don't they? I've been to two of Watson's hearings and one for Atkins's. Watson has two children now through conjugal visitations—this man who killed my pregnant daughter. Susan Atkins also had conjugal visitations when she was married to a man from Texas. She's divorced now, but I think she has a new boyfriend.

"They both have this wholesome image that they play because their ambition in life is to get out of prison; mine is to keep them in, and I'll betcha two bits I win."

"I want to show you more of my interview with Manson, and then get your reaction. Roll the tape, please," Rivera told the producer.

The lights dimmed to spotlight Manson with a haircut a child might have given him, tattoos from elbow to wrist, and fingernails long enough to make Barbra Streisand jealous. I studied him, not with venom or fear, but to gauge. Though I'm grateful for Vince Bugliosi's helter-skelter motive and the convictions it brought, I don't buy into it for a second. There's something more, some deeper motive for the killings. Even though Manson talks in riddles, he seldom lies. So I watch and wait for that morsel of truth that might slip from his lips, revealing the true motive. . . .

"Help me understand something, Charlie. . . . Why did those girls murder for you? Why did Tex Watson murder for you?"

"They didn't murder for me."

"You told them to."

"No, no, no. Come back, DA. Come back. That's not reality."

"What is?"

"Reality is, they did what they did. I didn't tell them to do nothing. They took it upon themselves to do what they did. They're responsible for their own actions. I'm responsible for my actions," Manson said, tapping on his chest.

"Let's be straight. What are you guilty of, then?"

"I'm guilty of thinking that I had rights in a courtroom. I thought I'd just stand in the courtroom and tell the judge the way it was and it'd be all over in fifteen minutes. . . .

"Consider, I tell you it's like this; yeah, I chopped up nine hogs, and I'm gonna chop up some more of you motherfuckers. I'm gonna kill as many of you as I can. I'm gonna pile you up to the sky. I figure about fifty million of you and I might be able to save my trees and my air and my water and my wildlife.". . .

"Why did you murder innocent people?"

"First of all, there were no innocent people to start with.". . .

Rivera's patience dwindled. "Why are you saying Sharon Tate wasn't innocent?"

"Wait a minute."

"No, I can't let you get away with that shit, Charlie."

"Terry Melcher was supposed to do some things with music that he didn't do. Terry Melcher lived in that house. Nobody knew Sharon Tate lived in that house. He broke his word, man."

"So you went looking for him?"

"No. See, you don't understand it. Let me lace it up your head again, man. When Melcher broke his word and didn't do what he said he was going to do; when Leno LaBianca got stabbed all up, and all that gold and stuff was laying around, and the little black phone book from the New York hit list was gone. . . . There's a lot more on this little road than you see. Dig?"

"No, Charlie, I don't. There are nine people dead and you've yet to cop to it."

"Son, there's a lot more than nine dead people out there."

"You want to make a confession today?"

"On what? I'm innocent, man.". . .

"Let's assume everybody misinterpreted what you're saying out at that ranch. And these kids took it upon themselves to do this deed. I'll play Tex Watson. . . . I come to you and say, 'I wanna go kill all those people.' "

"It wasn't like that. Tex come to me after a drug deal and says, 'The guy beat me for my money. What should I do?' I says, 'Whatever's in your heart.' So he goes and beats up this broad who helped steal his money and then takes off with the cash. Then he ran off and left me to face his responsibility. You dig? So, I go see this guy that Tex owes . . . and I end up shooting that dude. You dig?"

"Did you kill him?"

"No. So then, Tex come back around; now he owed me one.". . .

"So, you did one for Tex?"

Manson nodded. "And you got to pay the brother back.". . .

"So what happened? . . . What did Tex do?". . .

"Tex went crazy, man; he went out of his mind," Manson said, in a defeated voice. "Dumb shit brought us all down."

"Now what about Susan Atkins? She comes home to you with bloody hands."

"Yeah. She says, 'Charlie, look what I did for you. I just killed myself and I give you the world.' I said, 'You dumb fucking cunt; I had the world. You just put me back in jail again.' ". . .

"How did she react to what you told her?"

Manson waved him off. "People don't hear each other. They talk to each other, but very seldom do we communicate."

The stage lights came up again. Rivera moved to the center of the audience. "Doris, how do you feel about what Manson said?"

"I agree with him on one point. Those that killed Sharon are the only ones responsible for what they did. Oh, they'll say they take responsibility, only it's followed by 'but.' 'I take full responsibility, *but* I was on drugs.' Or 'I take full responsibility, *but* I was under Manson's control.' So it's a façade, you know. And, I might add, they have yet to say they're sorry for their actions. It's all me, me, me when they talk about the repercussions."

"What was Manson's responsibility?" Rivera asked.

"He sent them to both houses, and he tied up the LaBiancas. Now you tell me, what's the difference in tying them up and plunging the knife? None."

"Do you think the others would have killed if it hadn't been for Manson?"

"Absolutely."

"Would you like to talk to Manson?"

"Oh, I'd love to."

"What would you ask him?"

"I'd ask him why he sent them to the house because he knew Sharon was there. He can deny it, but he knew, and I want to know why."

"During my interview with him," Rivera said, "Manson told me that you don't go to his parole hearings. Is that true?"

"I've never felt the need to attend one of his hearings. I mean, you've seen how he jumps around, waving his arms and babbling; he does enough to keep himself in prison. If the day comes that I have to, I'll be there. Right now, I only attend the hearings of Watson and Atkins, the ones that actually killed Sharon."

Rivera put his hand on my shoulder. "You've told me how difficult it is to go to their parole hearings. How long will you continue to put yourself through that ordeal?"

"For as long as I live—and then some." I laughed.

"What's next for you?"

"Watson's got another parole hearing next month, that's as far in the future as I can look."

Patti

I sat with my parents in their family room, watching the *Geraldo Show*. The credits rolled at the end, and so did their home address as a contact for POMC.

I looked at Mom. "Have you lost your mind?"

"There are grieving parents out there who need my help. How else are they going to find me?"

"The post office box that you used for the petition drive!"

"You worry too much, you always have."

"And you don't worry enough."

"I am not going to get into an argument with you every time I do something public. Let's change the subject—not that the next one will be any better. They've delayed Watson's parole hearing to the week I'm in Washington. I'm scheduled to testify at Congress—"

"You have lost your mind, haven't you?"

"Patti, someone has to make an impact statement at Watson's hearing."

"Not me," I said.

"How can you turn your back—"

"I'll go." Dad's comment stunned us both to silence. We turned toward him, as if another being had invaded his body. He tapped his pipe free of the burned tobacco and repacked more. "You heard me, I'll go."

Mom's wits came to her. "P.J., you won't get a gun into that prison."

"I'm serious now," he insisted. "I'm tired of seeing that asshole preach the gospel like it means anything. I watched him eight hours a day during that trial with a Bible in front of him the whole time. His religion didn't make any difference then, and it means nothing now. I'd like to give him and the board a bit of advice—the Holy Ghost himself could show up to ask for Watson's parole, but the only release he should get is a ticket straight to hell."

Mom looked at him, waiting for the punch line. He put his right hand up for oath. "You have my word," he said, "I won't make a ruckus."

P.J.

I refused to sit through Watson's entire hearing. Instead, I waited in an airless chamber adjacent to the boardroom until it was time for my statement. I rubbed my temples. Damned parole hearings; I still didn't know what to say in there—oh, I had lots of things that I wanted to say, but they'd throw me out on my ass.

I couldn't think of anything to say because I wasn't here to give anyone a piece of my mind. I was here to save Patti the experience, and because I loved my wife enough to support her efforts.

Doris and I never did cross paths on our grieving trails. She'd crawled her way out of hell and left me behind, where I wallow in the gloom and bathe in the heat of hatred. I'd never admit it, but I was damn proud of her work, and damn jealous that she'd found residence above the licking flames of revenge.

Oh, I wasn't any happier about all the press, but I'd taken Doris's suggestion to heart and kept quiet. When journalists came to the house for an interview, I'd stay long enough to leer at the interloper who'd ventured over the threshold, then disappear into my study until they left. Though I'd never forgive the press for the lies about Sharon, I'd moved on to a fresh resentment. With the cameras rolling, they probed my wife until they got the tearful response that would boost their ratings. Each one of those interviews ripped another piece of her soul away, and I feared that in time, there'd be nothing left.

I toyed with the idea of Watson's demise. An anonymous phone call came each year on the anniversary of the murders. The first year the caller explained in detail his plan to assassinate each one of the killers; in subsequent years I'd pick up the phone to hear the voice, "Paul, it's that time of year. Just checking in."

The unknown man had instructed me to answer only with a yes or no. So far my answer was no. Right now, I wished I'd said yes. I paced a mile's worth of resentment, yet each step backtracked me eighteen years into the past when I'd waited much like this to testify at the trials to ensure the killers a gas-chambered vacation. With their deaths signed, sealed, and delivered by the judge, I'd expected only to encounter the killers again in that stainless-steel green room where the state dropped the pill, and I'd have the pleasure of watching the life drain from their eyes. The news that the supreme court overturned those sentences had sucker punched me a good one that I'd yet to straighten out of.

I stretched near the barred window and looked at my watch. An hour had passed, and I still didn't know what to say to the parole officers. The door swung open. "Come on, Mr. Tate, they're ready."

In a cramped room, I sat rigidly in the chair catty-corner to Watson's position at the table. I stared at him with suppressed though fermenting memories.

During Watson's trial, I spent hours watching him perform to complement his insanity plea. All college-boyed out, with a dazed expression and a mouth gaping with the threat of drooling, he parodied an innocent, victimized by drugs and Manson. But then the day of reckoning arrived. Bugliosi cross-examined Watson, compounding pressure on him with each question until the killer's alter ego appeared. If I had sneezed, I would have missed it, but there it was; Watson's eyes narrowed into pure evil—a skin this snake would never shed.

I looked Watson over. Nothing much had changed, save a few wrinkles around his eyes. I took a beat to study the faces of the three men deciding his fate. Could they comprehend the grief that Watson's actions had caused? Did they understand the hell of these hearings year after year? Somehow, I had to influence them.

When I finally started, it surprised me how easily the words slipped from my heart. "I don't believe that there is anyone in this room who saw the crime scene as I did. I personally saw what Watson did to my child. I saw it as the bodies were taken out, and I saw it caked with blood. And, gentlemen, let me tell you something, there are no words to describe it, and no photos that do it justice, but it's right here," I tapped at my temple, "vividly etched in my memory, because I had to clean up what Mr. Watson left behind. So I am totally aware of everything as it was, and that man should never, never, never be turned loose on society to re-create such a scene. I've been watching him for eighteen years; he hasn't changed one iota.

"At past hearings, Watson has relayed to you what it's like for him to have a family, and family visits. I've got Sharon's bloodstained trunks in my home, and I see my daughter's blood on those trunks—those are my family visits.

"My wife told me at the last hearing he asked to be released so that he could see his mom and dad before they died. But you've already given them [the] greatest gift any parent could receive; they can still visit their child. I have to live with the fact that I will never, never see my child again because of what Tex Watson has done.

"Now, one last thought. Mr. Kay informed me that Watson has the magic number of one hundred letters that say we should turn him loose. If that has any bearing on the board, I'll get you a million letters. My wife can go out and speak and the letters will come pouring in—the letters are still coming in from the last time. And they all agree, that man sitting there should not be turned loose on society. I have nothing more to add. Thank you."

The guard escorted me back to the same room to await the decision. I'd once seen Watson on a news report from a past hearing. He commented, "I have to live with it daily just as the families do. There's much bitterness on my part."

Watson had not a damned idea of what me or mine have had to live with, because if he had even an inkling, he'd have put a pistol to his head years ago instead of asking to be released.

Lap after lap my mind reeled. I clipped circles trying to keep up with my adrenaline. Had I'd made a positive impact? Had I said the right things? Did I sound too angry? What if the one time I attended a parole hearing, they decided to release one of Sharon's killers? The rattling doorknob pulled me from the thoughts. I looked, but didn't believe. I closed my eyes, thinking the apparition would fade away. When they opened again, Charles Watson stepped into the room. The ever-present smile on his face vanished midstride of his second step.

Time stopped.

Each damning fantasy I had pent up over the years danced in a kaleidoscope of nightmarish violence no sane man would imagine. The countless hours I'd sidled next to the devil, knocking around strategies, even deals to unleash unspeakable torture on Watson was at hand.

Just five seconds had passed and already a battle of wills ricocheted in my mind, one side repeating the promise to Doris, the other urging retribution. Promises or not, I had rehearsed, defined, and then refined the details of just such a moment ad nauseam. Like a famished tiger, I salivated over Watson's neck, feeling it snap. With a will of their own, my legs took a step that pulled him closer to my grasp. My heart pumped, thumped, and then pounded in my ears as I took another step. Watson recoiled, his eyes glued to mine. Before the third step, a voice stopped me. "Daddy, don't do this."

Like brakes pounded at the last second, I skidded, swerved, and deflected a head-on collision with a nasty bit of fate. I may have found a cozy home perched on that fiery lounger, a happy disciple of the devil, but I wouldn't bring my family down with me.

My hands clenched into tight fists while my body shuddered from unreleased anger. I took another step, close enough to smell the widening stains under Watson's arms. "Boy, you had better hope that you never walk out of this prison a free man, because I promise you, there will be hell to pay on the other side. You won't know where or when, but one day, you'll look over your shoulder and I will be there.

"You probably don't know this about me, but I'm an equal-opportunity kind of guy, and I believe you should die the same way you killed my baby—if I tack on some interest for this long overdue debt, you'll wish the gas chamber had gotten you after all."

Watson didn't twinge a muscle for fear that even the slightest movement might be provoking.

The spell of tension broke with a prison guard's entrance. My eyes simmered on Watson. "Guard, I believe you've brought this boy to the wrong room."

Watson backed into the hallway without losing eye contact until the door closed to a safety barrier.

Sinking into the chair, I thought about the voice. Had it been Sharon's, or my subconscious reminding me that she wouldn't have wanted an outbreak of violence? In either case, she saved both our asses.

I pulled a cigar from my pocket before my arms folded in satisfaction. As surely as I'd never forget hearing Sharon's voice that day, I would never forget the fear I'd created in Watson's eyes. It was a look I'd waited many years to see, and one to savor for many to come. Like a drug addict, I already wanted more.

The army trained me on the importance of control for winning any battle; the next time there might not be a voice to stop me. My fury still ran too deep and that made me unpredictable. I watched Watson through the door window, knowing it would be the last time I'd be able to see him face-to-face, unless he was released.

12

A WOLF IN SHEEP'S CLOTHING

I'm not going to judge whether Watson is born again;
that's between him and his god. But I will tell you this,
there is no relationship between faith and release.

—DORIS TATE

Doris   1988

Along the California coast, midway between the metropolises of Los Angeles and San Francisco, sits the city of San Luis Obispo.

Sculpted hills, interrupted by ancient volcano remains jutting from the valley floor, surround the ten-square-mile community where the citizens live the SLO life.

Within miles of the settlement's heart, one can find the Hearst Castle, Cuesta College, and the California Men's Colony Prison. Downtown, along with the locals, a creek meanders through the tree-lined streets of historical buildings, including twenty-eight churches.

At the Mazarine Church, a group of day-care children romped around the playground. Under the shade of an elm, Karla watched over the children, and in particular, two boys—one towhead, one strawberry—as they bounced on the teeter-totter next to her. Their conversation caused her to set her sandwich aside.

"What does your dad do?" asked the redhead.

"He works at the prison. He's a minister."

"You mean he's *in* prison. My mom told me he's a monster."

"He is not."

"Is too!"

"Na-huh."

"Is too, is too, is too!" The boy jumped off the high point of the board. "I'm not even supposed to play with you," he said, and ran off to join the kids on the swings.

That truth could be stranger than fiction certainly ran through Karla's mind as she went to the lonely Alex Watson, whose familiar name had never crossed her mind until this moment. "Alex, what's your daddy's name?"

"Charles Denton—and he's not a monster! He's just my dad," he said, and then sprinted away, surely to hide his tears. Her heart went out to his innocence, but. . . .

Karla leaned into the family van. "Mrs. Watson?"

"Call me Kristin," said the driver.

"I won't be able to take care of Alex anymore."

The woman looked defeated. "I know what you're thinking, but Charles has devoted his life to Christ now. He walks with the Lord, not with Manson."

Karla held up a hand to stop her speech. "I'm a relative of Sharon Tate's."

"I'm really sorry about Sharon, but all of Charles's victims need to follow the Lord's example of forgiveness."

Karla considered her point for a moment, and then said, "It's hard to forgive someone who's never apologized nor asked for forgiveness. It seems to me that if he were truly walking with the Lord, he would have made that his priority."

"If Charles tried that, he'd only have the door slammed in his face because your family has fallen off the path with your anger and re-venge."

Their conversation turned into a heated debate, calling attention from those around them. Karla moved away from the van, then turned with a last thought. "If he doesn't try to open the door, he'll never know the outcome."

Karla didn't tell me about the incident, so the letter I received from Watson a few weeks later came as a surprise.

> *Dear Mrs. Tate:*
>
> *So many things have happened to bring us together over the last few months, that it is hard to refute that it is from the Lord. It is not that I have not wanted to write you for years. Up until recently, I didn't know that you were concerned about why I had not written to make a personal apology to you. I have been fearful in some ways of not knowing how you would take my contacting you. I would not want to cause you even more agony than I have already.*
>
> *After I became a Christian in 1975, this apology was a major concern of my heart. I talked with chaplains and attorneys about contacting all the families of my victims in hopes to bring us together, but nothing happened. . . . Regardless, there is no excuse for me not adequately expressing my deep regret and remorse to you in the boardroom. . . .*
>
> *Over the last few years, a greater and greater realization of what I have done impacts upon my heart. I pray that someday our hearts can come together in a deep personal way to share our true feelings with one another. I envision us doing this on a visit sometime, which I pray will eventually come. . . .*
>
> *I realize all this is probably too early for us, but several things have taken place to bring us together. Several people who are your friends have contacted me in a very positive way, praying that one day that healing will come.*
>
> *A few Sundays ago, my wife, Kristin, found out that her acquaintance was [a relative of yours]. It was a shock to know that your [relative] had baby-sat for my son. . . .*

Kristin had a great talk with [her], which gave me the go-ahead to write this letter. . . . I know that I have hurt you in ways that I cannot even imagine. . . . I pray that one day you can personally share your pain with me, and I can share my deep sorrow with you. I experience daily my hurts as my tormentors surround me. I do not expect you to let up on me in any way. I deserve all your resentment and anger.

We will never forget what has happened, but it is my prayer that we may come into a more complete forgiveness in the face of Christ. I pray you will give me a chance, not a second one in life, but one to share my heart with you. I pray you understand my hesitance in not contacting you until now, and that God will touch your heart with his sufficient grace to forbear all that is to come to pass.

Humbly in Christ our Lord, Charles D. Watson

I crumpled the letter, but then something occurred to me. I must have overlooked it. I reread it. Five pages of drivel, but Watson hadn't apologized. Like everything he did, the letter was a ruse.

I popped two aspirin over my greater concern, the supposed conversation Karla had with Kristin Watson. Karla was a born-again Christian, too, and there was no telling what she may have said. I put the letter aside and picked up the phone.

Not only did Karla confirm the conversation, but after thinking it over, she decided that Watson deserved a second chance because he now "walked in the grace of God."

"You're being played for a fool by him, Karla. He's just using Christianity as a ploy to get out of prison," I said.

"How do you know? You're bringing judgment against him without any proof."

"Proof? Killers like Watson don't find God; they find ways to get out of prison."

"That's your assumption," Karla countered. "As a Christian you must give him the benefit of the doubt, or you're no better person than you think he is."

Sure that my momentary loss for words wouldn't last, I slammed the phone down before something horrible oozed out.

How could my own flesh and blood be in support of Watson? I paced from room to room like a dislodged live wire looking for a source to release its energy. I was damn good and tired of Watson selling himself as the golden boy. He'd gone from awaiting execution to being housed in the country club of prisons. He enjoyed conjugal visits with his wife, which afforded him a family. He received state-granted educational, medical, and psychiatric help for him *and his family*. From his carpeted cell, he operated a business, disguised as a nonprofit ministry, which under the shelter of donations netted $1,200 per month, tax-free. For lack of a better phrase, he'd been getting away with murder long enough.

If people wanted proof, then proof was what I'd find.

In my heart, I knew that Watson was a wolf in sheep's clothing who used his religion, ministry, family, and supposed remorse for his crimes as a front.

Exposing Watson's true agenda, and substantiating that he was at present no more rehabilitated than the day he'd entered the prison system, became my obsession. And there would be no peace of mind until I corroborated that assessment.

A few years back, I watched an interview with Charles Manson in which the reporter asked, "Tex has conjugal visitations, he's fathered children, but you can't even have physical contact with visitors. How do you feel about that?"

In a rare lucid moment, Manson rubbed his beard then quietly replied, "I guess Tex is smarter than I am. He's got a college education. Tex knows you got to be smart in jail to get out."

Watson was smart. Nevertheless, over an eighteen-year period even the most cunning of liars trips over his pretexts. Determined to find those discrepancies, I planned to chronicle everything Watson had said or written regarding his crimes and rehabilitation. From his

1971 trial to his recent ministry testimonials, I searched to find the hidden Waldo.

WATSON'S MURDER TRIAL began on August 14, 1971. Without the chaotic support of the Manson Family, and therefore the press coverage, Tex melted in with the rest of the anonymous killers crossing through the halls of justice. Watson, however, had a disadvantage over the others accused by the state. Vincent Bugliosi was his prosecutor.

During the Manson trial, Bugliosi introduced an exhaustive amount of evidence to prove Manson's domination over his followers and therefore their willingness to kill on his command. Since Watson had pled not guilty by reason of insanity, the prosecutor presented evidence that Watson was a freethinking individual with a full grasp of his mental faculties.

Defense attorney Sam Bubrick called upon eight psychiatrists to testify that his client was insane.

Although he denied stabbing Sharon, Watson followed the doctors to the witness stand, where he conceded to participating in the murders, but solely for the purpose of pinning any calculating malice, aforethought, or acts to avoid detection on Manson, Atkins, Krenwinkel, and Kasabian.

Watson's confession came with other stipulations. One, he wanted it known that he fell completely under Manson's control—so much so that he thought Manson was Christ. Two, Watson wanted it specified in his testimony that on the date of the murders he had taken a multitude of mentally incapacitating drugs.

But strand by strand, during the cross-examination, Bugliosi wove a rope that left Watson a dangling, legally sane perjurer. "I believe you said in November of 1968 you left Manson. Is that correct?" Bugliosi asked at the trial.

"Yes," Watson said.

"Did you tell Charlie you were leaving?"

"No, I did not."

"Then after that two-month period you got in touch with Charlie by calling him at the ranch?"

"Yes. And at that point, Manson convinced me to come out and just visit. His voice was so hypnotic I just had to go back."

Watson's rebounds to Manson had nothing to do with mystical powers. Watson returned only during intervals of failure—after running out of money, drug arrests, or encountering trouble from ripping off drug dealers. In reality, Watson spent a minimal amount of time at Spahn's Ranch, leaving Manson on too tight a schedule to possibly program Tex with the Helter Skelter/second coming of Christ dogma.

"Tex, you have spoken to many, many psychiatrists since your incarceration, isn't that right?" Bugliosi questioned.

"Yes, that is correct."

"And you went into great detail about Manson, and your relationship with him; isn't that correct?"

"Yes, I did."

"Isn't it true, Tex, that you never told one single psychiatrist that you thought Charles Manson was Jesus Christ?"

"I might have used the word *supreme being* or *Messiah*."

Bugliosi knew better. In nearly two hundred pages of psychiatric reports, Watson *never* referenced that very cornerstone to his diminished capacity argument.

The question of whether the killers had taken drugs on the night of the murders was a point of contention with me because it mitigated their culpability.

During a talk show that I paneled with Vince Bugliosi, I asked him why he was so positive the killers were drug-free the night they killed Sharon. Vince said, "It's common investigative belief that when we interview a witness or a suspect, that their initial response to questions is usually the most truthful information we'll ever get out of them. This initial statement is made before they've had a chance to talk with accomplices, rehearse a story, read other accounts, or had time to contemplate another theory. Now, Susan Atkins's initial statement to me—before any of the other killers were even arrested—was that they had not taken any drugs. She said Manson specifically instructed them to have clear minds when they murdered. There's no doubt, and I proved it at the trials."

On cross-examination, Bugliosi asked, "Tex, on the night of the Tate murders you claim to have had belladonna, speed, and LSD in your system. Is that correct?"

"Yes, that would be correct."

"Did you have anything else in your system?"

"Nope."

"Do you recall telling a doctor by the name of Dr. Frank that you also took cocaine?"

"No, I do not recall that."

"Do you recall telling another doctor, Dr. Bohr, that the only drug you took the night of the Tate murders was belladonna?"

"No, I don't recall that."

"Now, on the previous night [before the murders] that would be Thursday. Did you take speed on Thursday night?"

"Yes. I stayed up all night, I believe."

"Did you see Charlie at all that night?"

"Yes, he was at the waterfall sleeping, and I had heard he was on belladonna."

"Did you talk to Charlie that night?"

"I really don't recall that much."

"But you saw him sleeping?"

"Yes."

"Just so we don't have any confusion here Tex, August seventh, that's a Thursday, August eighth is a Friday [the night Sharon was murdered], August ninth, a Saturday. You took speed on August seventh, and you were up all night, is that correct?"

"Yes."

"So, you were up in the early-morning hours of Friday, August eight, at the waterfall, is that correct?"

"Yes, that is correct."

"You also took some belladonna on August eighth, a Friday?"

"Yes."

"And you say you saw Charlie sleeping?"

"Yes, I did."

"What would you say, Tex, if I told you Charles Manson wasn't even in Los Angeles on the morning of August eighth?"

"Um."

"Tex, why don't you admit to these folks on the jury that you had no drugs in your system on the night of the Tate murders?"

"Objection!" Bubrick said.

Bugliosi continued without missing a beat. "You lied about Manson and you're lying about the drugs."

Watson's testimony to the amount of drugs he had consumed on the night he murdered Sharon was a ridiculous notion. Had he been on belladonna, LSD, cocaine, and speed he would barely have been able to walk, let alone climb a telephone pole (with its first rung six feet from the ground) while carrying thirty-pound bolt cutters, scale the seven-foot fence surrounding the estate, and still have enough wit about him to murder five people.

By the time Watson composed his autobiography, this must have been apparent to him as well. In his book, *Will You Die for Me?* he wrote, "While Manson went back to the movie set to round up Sadie [Atkins], Katie [Krenwinkel] and Linda [Kasabian], I reeled over the porch where Sadie and I kept our jar of speed hidden. I took a couple of deep snorts."

The reason Watson continues to claim that he was on speed is best said by him. "I made the choices that I made, but the methamphetamine caused all the violence, all the anger, all the hatred. It caused the rebellion and the resentment. Everything that was in me that I thought the drugs were taking away from me, were actually bringing me to that state of nothingness, and that raging individual was ready to come out."

Bugliosi had a few more strands to gather before his rope would be a readied noose. First, he needed to prove that Watson employed premeditation and deliberation. Second, he had to prove that Watson could decipher reality, as well as right from wrong.

"Is it your testimony, Tex, that on the night of the Tate murders you were doing whatever the girls told you to do?" Bugliosi asked at the trial.

"I was doing what Charles Manson had told me to do, and Charles Manson was the girls, and I was Charles Manson, and we were all Charles Manson."

"Uh-huh. On the night of the Tate murders you knew that you were going to Terry Melcher's former residence to kill everyone inside, right?"

"I really had no thought of what even murder was. I was just doing what Charlie and the girls told me to do."

"You knew you weren't going to the Tate residence to play canasta didn't you, Tex?" Bugliosi mocked.

"I had no thought."

"Have you ever heard of canasta, Tex?"

"Yes, I have."

"Have you ever heard of volleyball?"

"Yes, I have."

"You weren't going there to do those things were you?"

"I had no thought of what I was doing."

"What was the knife in your hand for?"

"It was put there by Mr. Manson."

"What were you going to do with it?"

"I was told to kill everybody in the house with it."

"Now the word *kill* comes out, not canasta, right, Tex? Kill? Isn't it true that no matter how many people were inside that residence you were going to kill them?"

"I was told by Mr. Manson to make sure everybody was dead."

"I am talking about after you were told by Mr. Manson, it was *your* state of mind that no matter who was inside that residence you were going to kill them. Right?"

"That's what they told me to do."

"What would you have done, Tex, if you arrived at the Tate residence and saw a squad of police cars? What would you have done then?"

"That's objectionable, Your Honor," Bubrick said.

"How did you know what to do if you never had any thought in your mind, Tex?"

"Well, I was being run by Mr. Manson."

Hogwash. I discovered an interview in which Watson contradicted this alibi by admitting that he'd actually defied Manson's instruction to kill all of Sharon's neighbors. "It [the murders at Cielo] was horrific and no way were we going to go to a second and third house even though Manson had ordered it."

In Watson's book, he wrote of the night he murdered Sharon, "We gathered up our clothes and weapons and quietly slipped back up the driveway. I carried the white rope over my shoulder. . . . I wrapped the rope around Sebring's neck and then slung it up over one of the rafters. . . . I started to tie the rope around Sharon's neck."

From the evidence table, Bugliosi picked up a bloodied piece of rope. "Tex, this is the rope that was tied around Sharon Tate's neck. Are you saying that you've never seen this rope before?"

"That is correct."

"You have no knowledge of how Sharon Tate had a rope tied around her neck?"

"No, I do not."

"Now, Tex, when you spoke with all those psychiatrists who interviewed you, did you lie to any of them about anything?"

"No, I told them like it was."

"How come you never told any of those psychiatrist[s] that Manson ordered you to bring rope, cut the telephone wires, wash blood off, and throw the clothing away?"

"I thought I told them that. Like I say, I don't know what I told them exactly."

"Isn't it true, Tex, that the only thing that Manson told you to do was go up there [to Cielo] and kill these people. And that it was *your* idea to cut the telephone wires?"

"No."

"And to bring the rope."

"No, that is not true."

"Now you claim that you were in the backseat of the car and Linda drove. Did you give Linda directions on how to get to the Cielo address?"

"No, I did not."

"Do you have any idea how she found her way there, Tex?"

"No, I do not."

"Linda Kasabian testified that while *you* drove to the Tate residence that you told her, Katie, and Sadie that you had been to the residence before, that you knew the layout, and that they were to do everything you told them to do. Do you deny that?"

"Yes, I do."

"You had been to the Cielo address about five times?"

"About three that I can remember."

"Then you know that there are a lot of trees and high bushes preventing a person from seeing the residence from the telephone pole. Isn't that right?"

"I think you are right."

"When the girls, as you claim, told you to cut the telephone wires, didn't it strike you as rather strange that they would know which telephone pole had wires that led to the Tate residence when you can't even see the house from the telephone pole?"

"I never asked any questions; I just climbed the pole."

"Well, isn't the real reason you didn't ask any questions, Tex, is that it was your idea to cut those telephone wires, and no one said 'boo' to you about doing it?"

"No, that is not correct."

If Sherlock Holmes had investigated Watson's guilt for premeditating Sharon's murder, he would have said, "There is nothing more deceptive than an obvious fact." If Manson ordered Watson to cut phone wires and to bring enough rope to tie up an army, why didn't Manson follow suit the next night when he led the killers to the La-Bianca house?

Unlike Manson, who had never entered the Cielo house, Watson had visited inside with Melcher several times, so he knew the living room had exposed beams he could hang a rope from, and he knew that a buzzer sounded in the house if the gate outside was activated. So only Watson knew to cut the gate communication wire. He even added the precaution of climbing the fence just in case he'd cut the wrong wire.

"Tex, I believe you testified yesterday that after you climbed over the front gate, a car approached, is that right?" Bugliosi asked on cross-examination.

"Yes, I remember seeing some headlights."

"And you went to the car and shot the man?"

"That is correct."

"Where did you shoot the man in the car?"

"I didn't see. I just shot at the thing that was there."

"Oh, the *thing* that was there. It was not a human being?"

"I didn't have any thought of human beings."

"Did the boy in the car have on glasses?"

"I didn't see his face."

"You testified yesterday that the people you murdered were like blobs to you. What do you mean by that?" Bugliosi asked.

"It was hard to see them," Watson said. "It was hard to tell what they were in a lot of ways, really."

"You knew they were human beings, didn't you?" Bugliosi stated.

"The thought of anything like that just didn't occur."

"Didn't you testify yesterday that the woman on the front lawn— number one, you're correct, it was a woman not an object—didn't you testify that she was covered with blood?"

"She was covered with blood, yes."

"Well, now, a woman with blood on her, that's not a blob, is it, Tex? Looked kind of like a woman with blood on her."

"Well, it's hard to say what she did look like."

"You also testified yesterday that there was a man inside. Again, not a blob, but a man and he was wearing blue jeans?"

"Right. That's right."

"Is that what you mean when you say 'blobs,' Tex? Men with blue jeans on and women with blood on them?"

"I didn't mean anything, you know."

"Did they beg you not to kill them; did they say, 'Please don't kill me? Please let me live'?"

"I couldn't hear that, no. I just heard a bunch of screams and hollers."

"Weren't you laughing with Diane Lake [a family member] while you told her how Sharon Tate pleaded for her life, and how much fun it was to kill her?"

"No. I did not."

"What did it feel like when you stabbed these people, Tex? What type of sensation was it?"

"I had no feeling."

"Did you see blood coming out of their bodies when you stabbed them?"

"I saw blood, but I don't know, I guess it was coming out."

"They were covered with blood, weren't they?"

"Yes."

"You knew they were human beings, didn't you."

"The thought of anything like that didn't occur. The only thing going through my head was just what Manson said."

Before penning his account of the murders, Watson noted for the reader: "It would actually be some time before I learned the names of our victims. That night, there were so many impersonal blobs to be dealt with as Charlie had instructed."

Two pages later, Watson wrote: "A terrified teenage boy [Steven Parent] looked up at me, his glasses flashing. As I lunged forward the boy cried out, 'Please don't hurt me. I'm your friend. I won't tell anyone I saw you here.'... Sebring turned back, protesting my roughness [toward Sharon] so I shot him.... He slumped over, still alive, breathing hard, groaning.... Sadie was sitting next to Sharon on the couch as the beautiful, pathetic, blond woman sobbed, begging us to take her with us, and let her have her baby before we killed her. It was the first time I'd realized she was pregnant. . . . My hand struck out over and over until the cries of 'Mother, Mother' stopped. . . ."

Watson also recalled that Woytek was "enormously powerful" and that Woytek screamed, "Oh God, help me, oh God." Watson even detailed the inflection in Abigail's voice: "As she lay on her back, she whispered, without emotion, 'I give up, you've got me.' "

Call me a crazy old lady if you will, but his memory is darned explicit regarding those blobs.

Years after Watson's book was published, he attempted to show he had compassion; in the process, he inadvertently made an admission. "The girls and I didn't enjoy murdering human beings, it was insanely difficult for us all, but our slavish hearts were committed. We wanted this outbreak of violence to be over with. We wanted to get the job done and leave. It was horrific for us."

Though Tex wouldn't admit it on the witness stand, the prosecutor had evidence that Watson fled Los Angeles immediately after the murders to avoid detection. He traveled first to Death Valley, Mexico, Hawaii, and then back home to Texas.

"Each day it seemed as though I got more confused," Watson wrote of his Texas homecoming. "Added to all the turmoil that had been boiling in my mind, it was obvious to everyone that something had happened to me in California. I didn't wash. I'd just lay around, watching television blindly with the shades drawn, screaming at my parents to shut up if they tried to speak to me."

Upon his arrival in Texas, Watson rekindled a high school romance. Since Denise had known Watson through the years—before and after Manson—her testimony was the most insightful of the entire trial.

Bugliosi held a picture at the witness stand. "Denise, do you recognize this as being a photograph of Charles Tex Watson as he looked in November of 1969, when you saw him in Texas?"

"Yes."

"During this week that you saw him, how did he dress?"

"He was always very neat; had on slacks and a shirt most of the time."

"Did he look clean?"

"Yes, sir."

"Did you notice any difference in his walk during this period?"

"No, sir."

"Or in the manner in which he spoke to you?"

"No, sir."

"He seemed to be the same old Tex that you had always known. Is that right?"

"Yes."

"Did Tex discuss the ranch that he lived on in California?"

"Yes, he talked a lot about it."

"Did he say with whom he lived on this ranch?"

"He said there were quite a few women, maybe thirty girls, but just a couple of men."

"Did he say who was the leader of this group?"

"He said that he and one other person were the main people."

"Did he tell you that he met anyone in California who, in his opinion, was kind of a supernatural being or Christ?"

"No, he did not."

Watson's inaugural proclamation of rehabilitation commenced in his jail cell before his trial began.

Watson the novelist asserted, "Prosecutor Bugliosi would insist that my claim to feel remorse was untrue, but he was wrong. As much as my scarred conscience was capable of feeling anything at the time, I had genuine sorrow for what I had done, for the unspeakable pain I had caused both the victims and those who loved them. I felt more deeply than ever before the reality of what we'd done those nights."

"Tex, how do you feel about Charlie Manson at the present time?" Bugliosi asked in the courtroom.

"I feel that he was a kind of false god or something. A false prophet, as you would say."

"Do you feel he was an evil man?"

"Yes, I do."

"And when did you come to that conclusion?"

"Well, since I have been slowly getting back to my parents and writing them every day. I don't know for how many months, but I have slowly been getting back to what I think is right."

"I believe you testified that you don't feel the same way about killing at the present time as you did at the time of the killings?"

"Yes, that's correct."

"You testified that you had no feelings then, but you do now. Is that correct?"

"Yes. My feelings are gaining each day," Watson said.

"Again, talking about Dr. Frank, when he interviewed you, did you tell him, 'I saw a guy laying on the couch. He started coming at me and I shot him and then I stabbed him and stabbed him and stabbed him. People were running everywhere like chickens with their heads cut off. I had no feelings then or now. It just doesn't affect me like it does others.' Now, when you said that to Dr. Frank—I had no feelings then or now—what did you mean by that?"

"Well—uh," Watson stammered, "at the time he interviewed me, I must not have had any feelings."

In a chapter Watson titled "On Trial," Tex avows that he welcomed Christ back into his life with an experience that occurred just before Dr. Frank interviewed him. "Hour after hour, I'd turn the pages of the Bible. As I did, something else began to happen inside me . . . I began tasting the reality of what I had actually done during those two nights of blood. Suddenly they [his victims] were not nameless, impersonal things, not pigs, they were terrified men and women who had begged to be allowed to live. I began to see that even for guilt as gross as mine, a penalty had already been paid. A death penalty, carried by God Himself in His son Jesus. Slowly I began to see the power of God's love to overcome that death and destruction, to heal it, not just abstractly but immediately and specifically for me. Charlie's trip had been death, but this Jesus promised life. God didn't turn away from those two nights of butchery. He took all that anguish and horror. He took the guilt of my bloody hands, and that if I would let go of it, it could be nailed up and done away with. It seemed impossible, too good to be true, but the Bible said it. Something inside of me said it, too. There could even be light in my darkness."

Despite these profound feelings, Watson sat in court daily, with a Bible in front of him, and plotted out his insanity plea. He swore in a court of law and before God to tell the truth, then turned around to tell blatant lies—and threaten to take another life.

"Tex, one last question," Bugliosi said. "You were interviewed by a Dr. Alfred Owre?"

"That is correct."

"Do you recall telling Dr. Owre, in October of 1971, 'I could kill you right now, very easily'?"

"Uh, I told him . . . see, that's what he said I said, but uh, I did say, uh . . . it was relating to him what kind of mind I had at the time of the killings . . . that I could have killed anyone at that time. That is what I said."

"When talking to Dr. Owre, weren't you talking about the present? Did you not tell him, 'I could kill you *right now,* very easily'?"

"Well, I guess I could have said that. Yes."

Would a man who rejoiced in the knowledge that God had forgiven him for his sins gone on to threaten another life?

After reading the rest of the trial testimony, I probed for a deeper understanding of what made Watson tick.

Turning again to his autobiography, I examined a passage that Watson penned on acclimating to an existence behind bars, in particular, psychiatric therapy. "The prisoner is caught in a bind. Attendance to group therapy in whatever game the psychologist may come up with are imperative, because such participation is necessary for good board reports when you come up for parole. Thus, by participation, the prisoner is blackmailed into supporting a bureaucratic system that does little or nothing for him. Not surprisingly, cynicism is widespread among inmates, and even the newest and most naïve prisoner soon learns the particular jargon and poses that will get him a good report from his shrink. It didn't take me long to figure out how to play the game."

To classify an inmate's progress, the Department of Corrections biannually assigns a group of seven psychiatrists to interview and gauge the rehabilitation for each convict. The parole board considers the psychiatric counsel's assessment invaluable when they review a prisoner for release.

Watson's impeccable reports from 1972 to 1976 showed him to be an expert sportsman. By 1978 Watson became so commendable, that the counsel concluded: "In Mr. Watson's case, retention in the CDC—California Department of Corrections—will have to be on grounds other than psychiatric ones."

In a world where an inmate is sent to the isolation unit for jumping the lunch line, missing an appointment with the psychiatric counsel wasn't taken lightly within the CDC. In 1980 Watson added his own game rules by defying the CDC's mandate and authority; he skipped his session to spend time with a visitor.

Two years later, the counsel called Watson for another interview. By this time, Watson had thrown away the playbook, and the counsel's findings were the most unfavorable he'd received since his incarceration. One of the doctors noted Watson was "a walking time bomb." Another felt, "Mr. Watson has a very high degree of suppressed hostility . . . who has only superficially changed since 1969." The seven psychiatrists agreed on an overall diagnosis: "Antisocial personality with characteristics of the narcissistic and histrionic personality disorders. . . . The extremely high degree of repressed hostility in a man who has committed heinous and sadistic crimes must be confronted through therapy. Until that hostility decreases substantially, Counsel agrees Mr. Watson has a high potential for violence."

Due to my early departure from his 1982 parole hearing, I missed Watson's fifteen-minute justification for the counsel's findings: "I feel that I'm sort of misunderstood by the psychiatrists here due to my lack of participation with them. Not that I have rebelled against anything they're doing, it's just that I got away from them since 1979.

"Because I had missed my last evaluation, when I walked into the 1981 evaluation, it was a very hostile atmosphere because they felt that I was some kind of prima donna. So I feel that reflected part of their bad evaluation. As I look back on it, I find that it was one of the worst mistakes that I ever made in my life—besides the murders. But since that time, I've availed myself to them again. I really want to work with the psychiatrists here."

Watson's personal vendetta theory had a flaw. The counsel's reports were based on an interpretation of a computer score of his answers rather than on their personal opinion of his answers.

Following that episode, Watson was very careful and re-entered the game. He was always in some type of therapy, kept every appointment, and was sure to tell the parole board at each subsequent hearing how beneficial the psychiatrists were to his rehabilitation.

I began all of this research seeking to reveal Watson's genuine persona. From Watson's own words, I found clues from his earlier background that unmasked the killer.

Born on December 2, 1945, in Dallas, Texas, Watson enjoyed a self-described happy childhood supported by a loving family.

From Watson's youth through his teens, he excelled in every endeavor. By age six, he worked hard to help his father run their country store. By eight, he labored in the onion fields near his home. At ten, a local reporter columned Watson as an industrious child for gathering crawdads to sell to the local fishermen. In his early teens, the town of McKinney, Texas, named him "Future Farmer of America" for raising a prize-winning calf.

In high school, Watson continued on a positive course with above average accolades. As an "A" student, he was popular among his peers, who voted him "Campus Kid" three out of the four years. He outshone his teammates in all the sports that he participated in, bringing home numerous first-place ribbons and lettering in basketball, football, and track.

Religion was important to the youthful Watson. "God was very much a part of my world," Watson wrote. "Next to my older brother, God was probably one of my favorite people."

Every Sunday, he attended church as well as led devotions for the youth group.

Like most high school boys, Watson led the simple life of cruising for girls in souped-up cars and drinking beer on the weekends.

Upon high school graduation, Watson had saved enough money to buy a new car and a higher education. At college, he was comparable to eighty-five percent of all students experiencing their first taste of freedom; partying, joining fraternities, and in between the fun, studying just enough to get by.

By his own account, Watson had strong goals, identity, and direction in life. So what happened to this all-American boy? "Inside I was beginning to feel as if God and my mother had one thing in common," Watson wrote. "They both wanted to hold me down, to keep me from doing the things I wanted to. My parents' world of church and God and rules wasn't what I wanted. I was a success; I could handle my life without them or that pale-faced Jesus in church. I started thinking about getting out, finding a more larger, more exciting world."

I referred to the psychiatric counsel's deduction of Watson's narcissistic tendencies and the puzzle pieces fell into place. Watson's ego led him to the killing field.

Watson decided California was the answer to his problems. He noted of his first day in Los Angeles: "It was a long way from Texas and if freedom was what I'd been looking for, I was certain this was it."

Once in Los Angeles, Watson began the steady descent that led him to Manson.

After unsuccessfully trying to hold down numerous jobs, he gave up on the business world and began selling drugs. It was his first step to making up his own bylaws.

By the time Watson found the Manson Family, he thought they were the answer to his dreams; they had dropped out of society, lived without rules, rent bills, or work hours. They existed to please only themselves.

When Watson and the Family's lives intertwined, they resided freely in Beach Boys' drummer Dennis Wilson's estate. The park-like setting held riches beyond their dreams: a huge home, swimming pool, fancy cars, drugs, an endless cash flow from Wilson, and every extravagance available to Hollywood's elite. As long as they didn't have to work for it, Manson, Watson, and the others embraced this life.

Eventually Wilson evicted the group from his house. Manson moved his flock to Spahn's Ranch. Within twenty-four hours, they'd gone from the lap of luxury to flea-ridden, dilapidated buildings; their meals scavenged from trash Dumpsters.

When the establishment stopped giving, the Family started stealing by entering homes in the dead of night, while the unsuspecting occupants slept through the "creepy-crawl" missions.

As time went on, the Family began to resent the success that others were willing to work hard at to achieve. They wanted the good life, but on their terms.

Like a contagious disease, money and fame is everywhere in Los Angeles, but it was just beyond Manson's reach. He desperately wanted to be a rock-and-roll star; he wanted to be idolized. When the entertainment industry turned its back on him, he decided to turn against society as a whole and everyone, save the crew of killers, would suffer because of this rejection.

Watson, Atkins, Krenwinkel, Van Houten, and the others followed his lead. Not because of brainwashing or believing he was Christ, but because they felt his same emotions; it's what united them as a group in the first place. Manson simply nourished the egos that already festered in each one of them.

In the end, the bitterness turned to violence. "I had anger in my heart," Watson said, "and I released it during the crimes."

I connected the dots from the trial transcripts to the parole hearings to recent interviews, and the picture turned out to be a conniving, self-serving man who would do and say anything to be released from his punishment. He lied throughout his trial in an attempt to avoid the death penalty. He came up with one excuse after another throughout his book as to why he murdered. To this day at his parole hearings, he won't admit that he thought of his victims as human beings whose lives he stole. From Watson's own confessions, there was no doubt that he understood not only that killing was wrong, but also, if he were caught, he would be punished.

Proof.

In a prison ministry interview, Watson related his thoughts after the murders when he left Manson to go back to Texas. "At last I'd be totally my own man, totally free, without anyone telling me what to do. That was what it boiled down to. I didn't want anyone, ever again, to tell me what to do."

Watson wants freedom; he wanted it in 1969, and he wants it now. What would he do if he ever received that pardon? I intended to make sure that question was never answered.

These photos from our family album reflect the way we like to remember
Sharon, Doris, P.J., and Patti.

Sharon's early years

Sharon when she was just a few days old (P.J. went AWOL from the navy to be there) . . .

Posing for her first portrait . . .

Horsing around . . .

And giving P.J. a big smile.

Sharon at home in Texas . . .

And in her grade-school classroom.

While P.J. was away at war, Sharon and Doris did everything together. They were very close to each other . . .

*And they were close
to Nannie Tate, too.*

*Throughout her life,
Sharon held family dear.*

Sharon all grown up

Sharon in her high school portrait . . .

As Miss Autorama in 1959 . . .

As Miss Richland that same year. (If P.J. hadn't transferred to Italy, she would have entered the Miss America pageant.)

Here she is with Jack Palance
as an extra on the set of
Barabbas . . .

With Pat Boone in Italy on the set of his
1961 ABC-TV special . . .

Basking in the sun on the set of the film
Don't Make Waves . . .

And in the spotlight with David
Niven while shooting the movie 13.

Sharon's first love was animals.

Meet Guinness, the first of many Yorkies Sharon and Doris would share. Doris later bred Guinness and gave a puppy from the litter to Sharon and Roman whom they named Dr. Sapirstein, a character from Rosemary's Baby.

Sharon's second love was the beach.

Her other loves were well known.
Here she is with Jay in Rome . . .

And later, with Roman in Santa Monica, California.

Patti throughout the years

*Patti really loved playing Barbies
with Sharon . . .*

*Mimicking her big sister, as she is playfully
doing here on their return trip from Italy . . .*

*Hanging out with Jay, who was like a
big brother to her, on his houseboat in
San Francisco . . .*

*And visiting Sharon on set. Here she is with
Tony Curtis during the filming of* Don't
Make Waves.

These are some favorite pictures of Patti and P.J. (Patti was expecting Brie at the time) . . .

P.J., Nannie Tate, and Patti.

Brie, Doris, and Patti—three generations of strong-willed Tate women . . .

And Doris and Patti on a mission!

Doris and P.J. through the years

P.J. and Doris in their youth . . .

And in their later years together.

Even after forty years of marriage, P.J. still liked to dance hand-in-hand with Doris.

After Doris died, P.J. would seek solace in his love of fishing.

He also found peace in the company of his grandchildren. Having been outnumbered by women for so long, his grandson Bryce was always a welcome presence.

Here's Bryce with Brie . . .

And several more pictures of Brie and P.J.
over the years.

Doris, of course, championed the rights of victims' families on every level she could, whether it was through filing petitions...

or impacting legislation (here she is with Governor Pete Wilson at the 1991 Victims' Rights March at the California state capital) ...

and while she worked tirelessly to help victims, her ultimate goal was to make the world a safer place for her family, especially her grandchildren (here with Brie).

*Shortly before she passed, Doris was honored by President Bush
for all she accomplished . . .*

*But no matter how tough a
person she appeared to be in
public, this is the Doris her
family knew and loved most.*

13

THE LOST SHEEP

Inmates at most prisons want to talk about how they've been victimized by the courts or the prison system or by society in general, almost as if their crimes never happened, their victims never existed. But if offenders here at Vacaville want to attend VORG then they have to leave their alibis and excuses out here on the prison yard because the one absolute requirement for getting in is that they own up to their crimes; no more denials, no more appeals.

—STONE PHILLIPS

Doris

On February 27, 1850, when murder was commonplace and there were more cattle than citizens, California legislature enacted the district attorney's office.

The few thousand settlers of Los Angeles County—dubbed Los Diablo for its lawlessness—elected William Ferrell as the first prosecutor for the people.

Along with Judge Agustin Olvera, Ferrell reinstated the City of Angels by holding court at the Bella Union Hotel, just a few blocks from where the Criminal Courts Building was eventually constructed.

On the ninth floor of the halls of justice, I crossed, uncrossed, and recrossed my legs as I waited in the reception area of the District Attorney's Office.

The current chief district attorney, Ira Reiner, and I were distantly at odds in 1971 during his short-term venture as Leslie Van Houten's defense lawyer. As the clock ticked past the second hour that he'd kept me waiting, I suspected he was avoiding that trip down memory lane. Be that as it may, I was on a mission, so I waited.

At the end of 1989 I accepted the California Department of Correction's invitation to serve as one of their advisory board members; in doing so, I became the liaison between the CDC and the victims. My job is to advise the CDC on how best to help victims through the judicial process—from the perpetrator's trial, to incarceration, and ultimately, the perpetrator's release.

The complaint most repeated by the victims is that they rarely knew the status of court proceedings against their assailants; therefore, they felt neglected by the prosecutors. Deputy District Attorney Cesar Sarmiento, from the Hardcore Gangs Division, solved this predicament by adding the victim or next-of-kin to the witness subpoena list, thus ensuring a victim's incorporation at all levels.

Intent to have Sarmiento's policy used throughout Los Angeles County, I sent Reiner my proposal and asked for his support.

I had reached a position where "no" was an unacceptable answer, and I'd grown accustomed to politicians from all over the state promptly returning my calls. Two weeks elapsed without a response from Reiner concerning the memo. On the fifteenth day, without an appointment, I stubbornly waited in his office until he took a meeting.

A handsome man with gray hair that didn't match his youthful face approached. "Mrs. Tate, I'm Chief Deputy Gil Garcetti. How nice of you to visit our office." He pulled a chair next to mine. "I'm sorry you've had to wait so long."

I took both his hands, and though I wasn't feeling the least bit hospitable, I turned on the southern charm. "Oh, honey, that's all right. How long until Mr. Reiner will be free?"

"Well, he's probably not going to get a chance to meet with you today, but he's brought me up to speed on your proposal, and we're going to try to get something together as soon as our schedule clears."

Still smiling and holding his hands, I nodded toward my purse, which held only my wallet and car keys. "You're not going to make me pull out all one hundred of these affidavits from victims who feel neglected by this office, are you? What I've proposed is such a simple solution; you could dictate it to your secretary in five minutes, and have it out to your deputies by next week." Before he could reply, I pressured, "I do have your support on this, don't I?"

At first equivocal, Garcetti followed through with an election-winning grin. "Of course you do, Mrs. Tate."

"Now that wasn't so painful was it? Three minutes and I'm out the door." I patted his hand. "Next time, let's not make this such a formal thing. Okay?"

In the hall, I pressed the elevator button. I looked at my watch; three hours for three minutes and my whole day is shot. Why'd they have to make this so damn hard? I stepped into the elevator car that moved as slow as the judicial process. It reminded me that no matter how fast I tried to run, it was still going to happen one step at a time.

Since my appointment to the advisory board, Jim Rowland, the director of the CDC, had approached me on numerous occasions to become involved in a program called VORG, Victim Offender Reconciliation Group.

California launched VORG on July 6, 1988, at the California Medical Facility in direct response to the recommendations made by the Governor's Task Force on Victims' Rights. VORG's goal was to reduce crime by increasing awareness on the part of the offenders. In theory, a face-to-face introduction to a victim would challenge the criminal's sensitivity to the enormity of negative and enduring damage they breed with their actions.

The CDC didn't offer any added incentives for the volunteering inmates. Their only requirement for membership was their absolute willingness to break the "denial syndrome" and accept full responsibility for the actions that brought them to prison.

At the time of Rowland's initial invitation, I couldn't distinguish the difference between the serial/mass murderer and those who committed centralized crimes of passion, or the first time offenders that the CDC firmly believed could be rehabilitated. In essence, I discerned VORG's objective as doing the unthinkable; lobbying to free killers like Watson and Atkins. So far, I doubted anything would change that ingrained perspective.

Patti 1990

The billowing clouds hung low enough to graze the midcrests of the Santa Ynez Mountains while Mom and I strolled the grounds of the Santa Barbara Mission; just a ten-minute drive from my home.

The only way to get Mom's undivided attention was to get her out of her home, which was often busier than the national convention headquarters. And today I needed her attention.

I'd never been big on planning the future—who knew if it would really come?—but having three kids forced a change of perspective, a change that unfortunately came a hair too late.

My husband's career in professional sports ended by default when he reached a certain age, and so did the absurdly high income he, like most athletes, was paid. A series of subsequent failed business ventures ate away at our savings until even the crumbs had been licked up. We owed everyone, including the IRS, who threatened to take our last asset, our beach house.

The plan my husband and I came up with wasn't much of a plan, but we were out of options. I hated transferring my burdens to Mom and though she'd never admit it, if I moved home with three kids, we'd cramp her lifestyle. It had been a long time since she'd had teenagers living with her, let alone a three-, six-, and nine-year-old.

"You've been quiet as a mouse all morning. What's knocking around in there?" Mom asked, while patting my head.

"We're going to have to rent out the house over the summer to make ends meet," I started.

"Good! You'll stay with us," she said as if she'd known the question before I revealed it. "And before you start up with nonsense about being a burden on me, I've hardly gotten to spend any time with the kids because of all your traveling. Now, let's stop this useless worrying and enjoy our day together."

She took my hand and tugged me forward, just as she had done my entire life.

On the flip side of the Santa Barbara Mission's tour tickets, the still functioning parish noted their ambition: "Actively seeking justice for all people, and to foster a local and global spirit of reconciliation and peace."

Under the shade of the sprawling sycamores lay the mission's cemetery, filled with tombs of Chumash Indians, elaborate mausoleums, a haunted majordomo house, a towering crucifix, and a statue of Saint Francis of Assisi.

We made our way along the adobe wall to the church entrance at the foot of the burial grounds. Atop the arched doorframe, a set of three skull and crossbones seemed to sentinel those who passed from the funerary gardens. "What do you suppose those mean?" I asked.

Mom laughed, "I don't know, but it makes you kind of hesitant to go through, doesn't it?"

Without so much as an umbra's notice, a priest startled us. "I rather like to think of them as a warning of the brevity of life," he said. "Sorry for intruding. I couldn't help but overhear your conversation. I'm Father Tim."

"I'm Doris, and this is my daughter Patti."

The priest, dressed in a brown woolen robe, spoke with the eloquence of a man with twice his years. He looked at Mom curiously. "Pardon my stare. Aren't you Doris Tate?"

"That I am."

With a touch to his cross, he presumably expressed a swift internal prayer. "Such a tragedy that was. I lived in Los Angeles at the time. I don't think anyone who lived there will forget that summer."

"You'd think that, wouldn't you?" Mom said. "But many have forgotten, and that's why I have to do what I do."

"Yes. I've seen your newspaper interviews. May I be so bold as to offer you some guidance?"

"Oh, darlin', I can always use a bit of advice from above."

"In order to release your soul from the burdens I know you must have, you must find forgiveness in humanity."

Mom shook her head no. "You're asking me to betray Sharon by absolving them."

"You're betraying your own spirit by not forgiving. Trust me, your daughter would want you released from this albatross. It is the only way you will find peace."

"You don't understand, and until you've experienced what I have, you couldn't," Mom explained.

Father Tim pondered the quandary. "Did you happen to pass our statue of Saint Francis of Assisi?"

"Yes," she said.

"Personally, I enjoy his prayers immensely. Would you honor me by going home this evening and taking a few minutes to study him? He's much more convincing than I could ever hope to be." A distant bell rang. "I must go. Saint Francis's supplications will be the first post in mending your fences—you have my word."

Alone again, Mom rolled her eyes. "Why is everyone trying to heal me?" She waved playfully at the skulls. "You all best save me a ticket to hell, 'cause I'm incurable," she laughed.

Doris

Saint Francis of Assisi is the patron saint of animals and nature, and the spiritual figure of peace, love, and religious diversity.

Bullheaded as I might be, I never turned down God's assistance. I went home and studied the works of the young Francis Bernardone, who graced Italy eight hundred years ago. His prayers spoke volumes, and I'm sure that one in particular was Father Tim's reference point: "Lord, make me an instrument of your peace. Where there is hatred, let me sow love; where there is injury, pardon; where there is doubt, faith; where there is despair, hope; where there is darkness, light, and where there is sadness, joy. Oh Divine Master, grant that I may not so much seek to be consoled as to console; to be understood as to understand; to be loved as to love. For it is in giving that we receive; it is in pardoning that we are pardoned; and it is in dying that we are born to eternal life."

I durably endeavored to be a good Christian; therefore, I knew something had to give in the absolution department. I wrestled with it for weeks before yielding to the certitude that it's beyond my ability. Pardoning is God's domain. As an alternative compromise, I forgave Sharon's killers through His grace. But, within the laws of man, this forgiveness didn't lessen the killer's culpability or diminish my ambition to keep them in prison.

The day following that settlement, and an hour before I was due at a two-day conference at Corcoran Prison, I took a tour of the facility. Over the years, I'd explored nearly every prison in California, and by 1990, I mundanely viewed them as an accustomed workplace.

There were strict rules to my vocation that I needed to adhere to in order to gain access through the razor-edged confines. The guidelines prohibited me from carrying a purse, wearing denim, hats, wigs, dresses with hemlines above the knees, or a bra with an underwire—which, thankfully to date, no one had checked. I had to leave medications, aspirin, food, and gum behind. Security did allow one handkerchief, a clear change purse containing less than three dollars, and a car key. Inside the transparent bag, the prison required me to keep two forms of identification.

Prior to admission, I passed through three metal detectors. As a young mother, looking to the future, I would have never imagined myself at sixty-six years being searched by prison guards in order to work among hardened criminals. By preference, I'd expected to be living on a ranch somewhere back in Texas with grandchildren at my feet while I flipped their favored meal of pancakes.

The grandchildren had come, but while I'm here, they seemed a lifetime away. I wondered why fate chose me to be different from the millions of other normal grandmothers? What precious moment am I missing today while I'm isolated from my family? I would never know, because for now, I passed through corridors that celled inmate after inmate like zoo exhibits.

I shadowed my guardian escort, whose fatigue outfit included handcuffs, pepper spray, and a baton. His blond hair was shorn to the scalp, save the flat top above his freckled face. Though his delivery was authoritative and knowledgeable, it didn't entirely hide his Kentucky gentility as he opened the door for me. "We provide a variety of special housing programs here," Tom explained, "from the general population levels to the security-housing unit where we are now. SHU accommodates inmates who pose a threat to others, and those who require protection, such as Mr. Manson. They tried mainstreaming Charlie at Vacaville until some Hare Krishna tried to make a bonfire out of him back in '84. He's been segregated ever since. For his protection as well as yours, Mrs. Tate, we'll skip his housing unit."

"When was Manson transferred here?" I asked.

"Let's see, Charlie was brought here about a year and a half ago, in March of 1989."

I traced the officer's steps through a locked gate and into another wing. "Has Manson gotten into much trouble?"

"Ah, he's mostly all bark, but he's been known to throw his feces at the guards—which they don't take kindly to."

We entered an area of enclosed, roomlike cells. Halfway down, one of the inmates caught my attention by tapping on the glass of his window, motioning me over to his door. As I got closer and looked through, I saw that the stubble-faced man was naked. His cell was void except for a mattress on the floor and a combination urinal/sink; there were no pictures, books, or personal items that make each of us individuals. I thought of the wasted life in that cell. How can a human being withstand such despair?

When I got close enough to the glass to be able to hear him, he said, "It must be horrible living out there."

Here's a man who lived in an environment that held no evidence of his existence in this world except for his body, yet we were at odds, asking the same question of how the other could endure the life before them. I didn't know how to answer his question.

Sadness encroached as I thought of the universal reason that had brought us together. I tried to imagine a place without violence; a place where I could be at home right now enjoying my family and this man didn't need to be caged. As I stared into his contemplative eyes, I wondered if he could be helped.

"Come on, Mrs. Tate. You don't want to get too close to the men," Tom cautioned. "The area we're about to go into is where we do a majority of the inmates' psychological programming; some are one-on-one sessions, some are group sessions. . . ."

At various crossroads, I sensed that God cleverly slipped me enlightening wisdom. As we turned the corner, I felt the hair rise on the back of my neck with another pull of His guidance. Scripted above the archway was Saint Francis's prayer. I paused to read it again, and one phrase stood apart from the others: "Grant that I might seek to be consoled as to console, to be understood as to understand."

As the day passed, the sentiment clung to me like a child's cry for help. What if I could convert prisoners through VORG? What if I could save one person from being victimized by reforming the predators?

VORG'S maiden conference for Vacaville State Penitentiary was in less than twenty-four hours. Before leaving Corcoran, I called Jim Rowland. "I haven't decided for sure, but can you get my clearance through at Vacaville in case I decide to go to the VORG meeting tomorrow?"

"Sure, Doris, no problem. If you do go, you'll need to get there by eleven. The meeting starts at noon."

Later that evening, I rested on the overly firm mattress in the hotel, fretting over the next day's encounter with seventy-five prisoners in a room lacking boundaries. Seventy-five men who had created at least seventy-five victims. Men who had stolen security, and men who had stolen innocence. Men who made widows. And men who made mothers cry at their children's funerals. Yet men who would soon complete their sentence and be released back into society. Like a windmill, thoughts were spinning full-tilt until the sails blurred into a circle I couldn't separate.

I reached into the nightstand to get the Bible, and then looked up. "You've gotten me this far, you may as well take me the rest of the way."

I randomly opened the Bible. When I looked down, the chapter title bolted from the page, "The Lost Sheep, Mathew 18:12." Intrigued, I read on:

"What do you think? If a man owns a hundred sheep and one of them wanders away, will he not leave the ninety-nine on the hills and go to look for the one that wandered off? And if he finds it, I tell you the truth, he is happier about that one sheep than about the ninety-nine that did not wander off. In the same way your Father in heaven is not willing that any of these little ones should be lost.

"If your brother sins against you, go and show him his fault, just between the two of you. If he listens to you, you will have won your brother over.

"But if he will not listen, take one or two others along so that every matter may be established by the testimony of two or three witnesses. If he refuses to listen to them, tell it to the church; and if he refuses to listen even to the church, treat him as a pagan."

The passage calmed the windmill to a breezy swirl of change in how I distinguished criminals. The prisoners participating in VORG could not be seen as villainous adversaries, rather as men who'd gone astray, men who needed openhearted courage to help them make amends.

BEFORE ENTERING THE prison classroom for the VORG meeting, I looked for a familiar face to ease the anxiety. From the door, I noticed the group had divided; the victims assembled in one corner, the inmates in the opposite, each viewing the other with a combination of suspicion and perplexity.

Against my instinct of going to the respite of the victims' camp, I ventured to the inmates' turf. Feeling as vulnerable as a balloon carried into a thorny rose garden, I picked the meanest-looking one in the bunch and extended my hand. "Hello, my name is Doris Tate, my daughter Sharon was murdered by Manson and his gang."

After a moment's waver, he stood to greet me. Distinct from his towering presence was his gentle manner. "My name's Gerry. I'm real sorry about your daughter."

As Gerry and I talked about our experiences, the others gathered around, hesitantly at first, then with curiosity as they watched us chatting comfortably. Time passed and soon the words came more easily, and eventually, spurts of laughter escaped from our circle.

The border segregating the two cliques lowered by the end of the meeting as the atmosphere dissolved the speculative doubt into warmth and respect. Stereotypes were lifted to where they didn't fit into the room anymore and people paired off, honestly communicating their fears and feelings about what's currently happening in their lives. For the first time in their criminal career of injury, these inmates associated a name, face, and tragic story to crime; their spirits seemingly touched by the encounter.

I appreciated that with each new group my story would compel the men, but how would I hold their interest in the future months?

A RARE WAVE of humidity doused Palos Verdes. The open windows had a minimal effect in the family room, where I packed in twenty-five POMC members who grabbed anything that could serve as a fan. A father was midstory, "The day of my boy's funeral they caught the bastard. I watched his arrest on TV and a reporter asked him, 'Why'd you shoot Jaimie?' He looked at the camera and shrugged. 'I had a bad day.' I remember thinking, pal, define a bad day, I just buried my son."

"You know, I've been trying to figure out ways of making a bigger impact on the VORG prisoners," I said, "and it just hit me while listening to some of your experiences. What if we created a video that had all types of victims who explained to the inmates what happens in the aftermath of their crimes?"

Unexpressed opinions held until Ernie spoke up. "Listen, Doris, I don't think there's too many of us that support this VORG idea. You're helping these guys instead of making it tougher on them—like they deserve."

"Yeah," Mike joined in, "I don't know how you sit in a room full of murderers and rapists; seems to me like you're riding the fence of supporting them."

"You all are entitled to your opinion," I said. "But let me tell you, I've been at this for nineteen years, eight of them fighting for our rights, and it's a slow process. I'm not about to stop lobbying for tougher laws, but in the meantime I need to work within the limits of what our laws do allow and that's rehabilitation and release.

"Like it or not, the prisoners that I'm working with will be released. So let me ask you, Mike, is there anything you wouldn't do to bring Rachel back? Ernie, how about Linda? If by working with these guys I can keep one set of parents from having to join our group in the future, I'll put everything I've got into it."

Shanee meekly raised her hand. "I'd like to do that, you know, to try to save another mother from having to go through this."

"Me too," said Idell. "One less child murdered is one less of us." She used a *TV Guide* to fan her face. "And God knows, we don't have room for any more."

Despite the criticism that fell upon me from POMC and other victims organizations, I pushed forward with the video. With the CDC's approval and funding, I found five victims willing to share their ordeal for the inmates.

VIDEO IN HAND, I entered the VORG classroom. "Today's going to be different," I told the men. "We're going to watch a movie."

I left the content a mystery and hit the PLAY button. I positioned my chair in the front of the room, facing away from the television, toward the inmates, and watched their reactions.

"It was two in the morning," explained the man on the tape. "My wife and I heard something outside, then a car door quietly closing. I thought it was my son coming home late, but when I looked out the window, I saw my car rolling down the hill.

"When friends found out my car was stolen, they'd say, 'Well, you had good insurance right?' And I did, but this great insurance took me five months to get any action from them. In between that time, the experience almost cost me my marriage.

"I work the graveyard shift at the docks, and the buses don't run during that time. I had no car, no transportation, so on many nights, I walked eight miles to work and eight miles home. I was bitter and depressed. . . .

"Thieves who think that what they do is okay because insurance will pay us back should know it's not that easy."

A mother of two sons began the next segment. "It was 6:30 in the evening and I was getting ready to go to a friend's for dinner. The next thing I know, my six-year-old came running in and said, 'Mommy, Mommy, there's robbers in the house.' I thought he was kidding, but before I knew it, there was a man putting one hand over my mouth and the other holding a sawed-off shotgun to my head. He demanded money. I told him to take anything he wanted. I didn't care. I just didn't want him to shoot us. My son was hysterical, and he threatened to shoot my boy if he didn't stop crying.

"After I gave him all my jewelry, I told him about a new stereo in the living room. At that point he let go of me to find the stereo. I didn't know what was happening with my other son so I walked into the hall to look for him. Halfway down, I found the babysitter, but she didn't know where my older boy was.

"Time was running out at that point. The gunmen were unmasked so I assumed they weren't intent on leaving witnesses. I had to make a decision that no mother should have to make. Either continue down the hall to search for my older son, or try to save my younger son and the babysitter. I decided to grab the younger one and the babysitter and pushed them into the bathroom. I locked the door and told them to lay down in the tub. I hid in a corner and started praying. . . .

"I'd like to tell the prisoners that are watching this that the trauma is not over with the incident; it goes on forever. It affects family and friends. We're people just like you. I'm sure many of you have families; how would you feel if this happened to your kids? How would you feel if you had to choose which child you were going to save?"

A rape victim looked into the camera lens. "I feel like I am serving a life sentence. They tell me that rape is worse than murder because I have to live with what happened to me. They took away my innocence. . . .

"It's always there. There's always a reminder. I still have flashbacks. I look at myself as a survivor, but what these men don't realize is that I am still a victim. . . . My feelings go so much deeper than what appears on the surface. My inner being was violated. My soul and my heart were stolen. The scars will never go away. I want to get on with my life, I want to be whole again, but I don't know if that's possible," she concluded.

The final victim on the video is a young mother whose four-year-old daughter was shot during a domestic dispute. "My daughter . . . was playing out front when we heard a shot from across the street. . . . I picked her up when the shooting started again. I had her about a foot off the ground when her head fell back really hard. She'd been shot in the face. I started yelling for help, but I knew when I picked her up that she was dying. I held her in my arms and looked into her eyes. When they closed, I knew she was gone.

"I loved my baby. I will never get over her. She was always so happy. My shadow, that's what I called her. She would come up and say, 'Mommy give me a kiss,' and she'd make a loud smacking sound. Then she would say, 'Mommy, give me a hug,' and she would wrap her arms around my neck and give me a big hug. She would look me right in the face and say, 'I love you, Mommy.' I miss that. Oh, I miss it so much. . . .

"Losing your child hurts more than words can say. People think it's an emotional pain, but it's also a physical pain. I feel like someone is squeezing the life from me because I'm missing my baby. . . .

"In eleven years, [my daughter's] killers will be able to walk out the door and join society again. In eleven years, I will still be going to the cemetery in order to tell my baby 'I love you.' If it's over for them in eleven years that's not fair because it's never going to be over for me. I want them to see the photographs of my baby growing up. I want them to see her as she was laughing, playing, or taking a bath. This was a living, breathing human being, a laughing child, not just a picture in the newspapers. I would never want another mother to go through what I went through. It's a walking hell."

For the video's conclusion, I took heed to the young mother's wishes, and shared pictures of her daughter during her short-lived life, including a picture of her tiny white coffin.

Throughout the half-hour, I studied the prisoners. Some of the inmates appeared uncomfortable and fidgety; others sat as if in a trance, some shed silent tears.

Along with some new members, there were the inmates I knew: Joe, who was serving three years for robbing motels to support his drug habit; Donald, a gang member serving fifteen years for breaking into a couple's house, robbing and beating them; Rashad, serving a life sentence for a murder spree in San Francisco; and Craig, who had shot his roommate during an argument. I turned off the television. "Well, what do you think?"

"I listened to what the lady had to say, but the casket at the end is what got to me. I realized that I hadn't seen a casket in thirteen years."

I craned my neck to make eye contact with Craig in the last row. "It's final, isn't it, honey? Murder is not reversible. Once you've taken a life, that person's never coming back. What you witnessed today are not isolated instances of victimization, it is the universal cry of all victims."

An armed robber who'd gunned down a clerk joined the discussion. "Mrs. Tate, how do you deal with the loss? I mean, you have your memory and your pictures, but you don't have that individual no more."

"I can truthfully say that it took me ten years; ten years . . ." Try as I may, a persistent sob snuffed my response.

"It's okay, Mrs. Tate. Take your time." A once-perceived enemy handed me a tissue. "Here, I know they don't let you bring these."

I took the Kleenex and a calming breath. "It took me three years just to admit that Sharon wasn't around anymore. It took me ten years to recuperate. Once I was prepared to face up to Sharon's death, then I was able to go on to counsel other families. There are simply no words to describe the grief of losing my daughter to murder. The intensity of that grief is at the bottom of the pit. That's where my grieving lies."

One by one, the men unveiled what lay beneath their icy veneer until even the coldest of the group spoke out. "We use to go in homes at night. There'd usually be families. Some were sleeping. We didn't care. We'd just wake them up, tie them up, beat them down and then we'd leave. Sometimes, we did drive-by shootings. It really wasn't for drugs or money, it was just for hate."

"Hate for what, honey?" I asked, wanting to understand.

The inmate looked down to his hands and shrugged. "Hate for the things that happened to us when we was young, how people abused us, for the things we couldn't get when we was little, so we'd just go around and take what we could take, do what we could do."

"We all have bitterness," I told the group. "I have bitterness, you have bitterness, but now I've taken my bitterness and turned it into belief, belief that through VORG we can help each other. I have to believe that it's all worthwhile. Otherwise we should all pack up and go back to what we did before."

Still suspicious, a gang member challenged my motivation. "After what you've been through, how can you come here and talk with us?"

Knowing my answer was going to make or break a bond with this virtual child, I chose my words carefully. "I can talk to you because this is a voluntary program. You only get out of it what you put into it. I can separate the good guys from the bad guys. I think that's why we're all here. I know that I am not talking to the Richard Ramirezes, Randy Krafts, or the Charles Mansons of the world."

I walked over to the nineteen-year-old who'd posed the question. Asserting physical contact, I put my hand on his shoulders and leaned to his level. "I think that I'm talking to people who want to be rehabilitated. Am I right?"

He had no choice but to look into my eyes. "Yes, ma'am."

Discussions continued for another hour. The men talked freely in my presence because I didn't treat them like an expert in the field, nor did I judge their actions. I concentrated on projecting myself as a nurturer who cared about their welfare. "I truly feel that if there are fewer of you, then there are going to be fewer victims, and frankly that's what matters most. I would like to have one hundred percent of you never commit another crime, but if I can rehabilitate even one, or two, or three of you, I have saved one victim's life."

THE CREW FROM *20/20* followed me to some of the VORG meetings for a report they titled, "Mrs. Tate's Crusade." I was excited when Stone Phillips arrived at the house to tape a one-on-one chat—even P.J., who usually grunted a hello to most reporters before hiding in his bedroom, stayed to watch the interview.

After a quick sound check, Phillips said, "You couldn't have done this ten years ago, could you?"

"No way. Too much pain. Too much Denial. *Denial.* It took three years to say that she was murdered. Three years to just say, she's not around, she's, you know, not in Europe doing a movie."

"These are guys [the inmates] who have become experts at telling people what they want to hear. Don't you ever feel like you're wasting your breath?"

"Never. Never. Because out of sixty, one of them is going to be affected. One of them will have a child of their own, okay, and will relate to my loss."

"Do you think they care?"

"I truly believe that some of them do. I have to believe that it's all worthwhile. Let's reverse that, okay. If I could save Sharon by what I'm doing now, how hard would I work at it? There's nothing too monumental."

I caught sight of P.J. behind the camera. We were recording in the family room that he'd painstakingly remodeled after Sharon's murder trial ended. The project was probably the only thing that kept him sane back then.

As he stood there in the army's "at ease" stance, with his arms crossed, and a pipe clenched between his teeth, he watched over me as the protector he'd always been—except for, on that August night twenty-two years ago when, in his mind, he believes he failed us. For years, my obstinate husband had claimed indifference to my plight, even balked at it, but in that instant, his look said otherwise. It said, I believe in you.

We'd come a long way, he and I. Sure, there were days when I wanted to wring his neck. Days when respect for one another drifted to the wayside and the love we had went right along with it. But days like this one, when we paused long enough to take note of the vows we'd made almost fifty years earlier, were the days that helped us make it through the thick and thin of it all.

14

THE INFORMANT

Mr. Watson is very strong-willed. He's determined to do it his own way. He's been recommended by the parole board to do certain things; participate with mental health counseling and programming in that area. And he has steadfastly refused and he has substituted his own program; which mainly is working in the chapel, a religion that he has more or less developed himself.

—ROBERT CARTER, SUPERINTENDENT, BOARD OF PRISON TERMS

Doris

I reserved Sundays for enlightenment, not only from the sun in the backyard, but also from government policies, enactments and amendments to laws, and newsletters from victim and law enforcement organizations.

Early in 1989 I founded COVER—Coalition for Victims' Equal Rights. The logo on COVER's letterhead is an umbrella, symbolizing my goal to unite all victims' groups in order to have a greater legislative impact. Run as an informational clearinghouse, COVER provided support and information to elevate public awareness to the plight of crime victims. The alliance had barely gotten off the ground before an influx of mail started arriving from all over the country.

On the day of rest, I scanned the letters, then grouped them from crucial to junk. I was about to toss one into the CDC pile, when I noticed the front label, MRS. TATE, URGENTLY CONFIDENTIAL. I emptied the contents of the package to discover an inmate's personal prison file with a picture affixed to the top.

The photo showed a man of his late forties with Brylcreem-encrusted hair that could be any color beneath the oil. Additionally, the package contained the California Men's Colony chapel brochures, an interoffice memo to the associate warden of CMC, a San Luis Obispo newspaper clipping, and a letter.

> *Dear Mrs. Tate:*
>
> *Please don't throw this letter aside. You are my only hope. My name is Steven Trouse. I'm serving a thirty-year sentence for murder. While incarcerated here at CMC, I became a born-again Christian. Although I was hesitant to become involved with anyone from the Manson Family, I found that the only way to be a part of the Protestant chapel here was to befriend Charles Watson and Manson murderer Bruce Davis because they run the show. . . . What I didn't realize until later, was how much power the two of them had here. . . .*
>
> *I have been in this institution for six years, and I've never had a disciplinary action brought against me until the day I decided to stand up to Charles Watson. . . .*
>
> *As you will see by the information I have sent you, Watson is listed as a minister on all of the chapel literature. The Reverend Stanley McGuire, who is supposed to be the supervising chaplain here has actually turned all reins of power over to Watson. In violation of the rules, Watson and Davis preach, teach, and counsel other inmates without supervision. I guess that wouldn't be so bad if that was the end of the story. What they are really doing while unsupervised is instilling fear and guilt in the other prisoners, and forming a new Manson cult right in front of the staff members. . . .*

In prison, Watson has refined himself, his methods, and his understanding of cults through his religious studies. Here, he has limitless opportunities to experiment on the inmate population. He is proficient at motivation and directing human behavior in this setting. He has learned to manipulate people with Christ, and he intends to capitalize on this all the way to freedom. Make no mistake, that is his only goal, and it's working. . . . He has more privileges than you could ever imagine, with extra conjugal visits, holiday dinners in the chapel, and no work assignments. . . .

Watson looks at the chapel not as a house of God, but a place of power. He uses his sermons as personal attacks against inmates who criticize his control of the chapel, often taking Bible Scripture out of context and using it to support his fire-and-brimstone messages. . . .

During the Sunday morning service on February 15, Watson's sermon was all about television being a tool of Satan. The Scripture he quoted to support his theory didn't even remotely correlate. I stood up during the service and in a polite manner told him I thought he was mistaken in his thoughts.

Watson said to me, "Sit down and shut up or I will have you removed." When I wouldn't sit down, he instructed the visiting clergyman to "Haul that man out of here." And they did just that. . . .

By the next day, on Watson's recommendation, I had a disciplinary paper signed against me for the incident. This is against the law. No prisoner has authority over another inmate. . . .

The following week, I wrote my appeal, including everything on Watson's power in this prison. Because of that revealing information, they put me in solitary confinement to shut me up. . . .

I don't deserve this, and I want it off my record. It's all

*because of some powerful hold that Watson seems to have
over the authorities in this place. Even though I went to the
press with this, nothing has changed here. I believe that you
are the only one they will listen to. Help me.*

Respectfully, Steven Trouse

I read the letter then read it again, the entire time, my face beamed along with the sun. For almost eight years, I'd gone astray from my vow to change Watson's prison life, and this letter was the catalyst that could help realize that intent.

Within Trouse's package, I discovered all the documentation that substantiated his accusations. Chapel literature featured Watson's picture along with his title of minister. Beside Watson, Bruce Davis was listed as associate pastor. Trouse also included Rev. McGuire's response to the allegations, as well as all the filed appeals on the incident.

Tantamount to a smoking gun, Trouse's assortment of goodies included an interoffice memo to the associate warden at CMC from the chairman reviewing Trouse's allegations: "Due to our conversation about Charles Watson and his role in the operation of the Protestant Chapel, we feel that a full investigation is in order.

"The Executive Committee has talked to approximately 250 inmates and some of the workers and ex-workers of the chapel. The following is the feedback that we have received: Charles Watson uses state supplies to print his Abounding Love Ministries newsletter. Funds that inmates have donated through the tithes and trust withdrawals are used to purchase personal coffee, sugar and cream, a popcorn machine, a video player, and holiday dinners, just to name a few, for the 'in group,' within the chapel. Inmate Watson spends a minimum of eight hours per day working in the chapel. This is in direct violation within the CDC: 'No inmate or parolee will be assigned as a minister or religious counselor on a full-time basis in lieu of regular work and program assignment.'

"The inmate population expressed concern about the influence and control that Watson has, and that he can have inmates, at his will, written up for an adverse action. Many inmates either have quit or have been fired from the chapel due to Watson's desire at the time.

"Due to the seriousness of the problem, this issue should be put on the agenda for the warden's meeting next week."

Elated as I was, I put on my indignant hat usually worn for adversaries and called CDC director Jim Rowland. "Jim, it's Doris Tate. Sorry to bother you at home, but this couldn't wait."

"That's all right. What's on your mind?"

"What in the hell is going on up at the Men's Colony?"

"Now, Doris, calm down," Jim said. "Are you talking about all this hubbub with Charles Watson?"

"Of course I am. I've been telling you guys for years now that something isn't sitting right up there. You've got the Manson Family running the prison chapel for God's sake—no pun intended. How can they let Davis and Watson be together? If they were on parole, you and I both know they'd be in violation."

"Doris, we're looking into it right now. They apparently made an effort at keeping the Manson people separated, but they only have limited ability to do that. As to what's going on in the chapel? It's a gray area. We don't like to get in the way of religious freedom."

"We're talking about the opposite of religious freedom here, okay. I have documentation that proves that Watson is running the show up there, actually taking the inmate's religious freedom away from them. And let me tell you, that's not the end of the story."

"What type of documentation are we talking about?"

"Let's just say it's from a reliable source on the inside."

He chuckled, "I hope you're not talking about Steven Trouse? Come on, Doris, you can't believe him. The guy's a convicted killer."

"Don't patronize me on this. I've sat on the sidelines for too long now when it comes to Watson. This guy's got someone up there wrapped around his finger. He's running a business from his cell—which I might add is illegal. He gets extra conjugal visits, five recommendations for him to move to another prison—yet he stays, and now this business with the prison chapel. I'm not letting it go this time."

For added effect, I played one of my trump cards. "It's not just Trouse. I have an interprison memo recommending that this problem with Watson be put on the agenda for the warden's meeting. Now, I want your guarantee that this is going to happen or so help me, I'll call a press conference and the next time you hear my voice it'll be on the evening news."

"All right, all right, I'll get into it. But I'm telling you this is all being blown out of proportion."

"Maybe, but you owe it to me to check it out."

For weeks, the correspondence between Trouse and I flowed with tidbits of information from him followed by more questions from me. "Who's protecting Watson?"

"I don't know his name, but it's someone in the governor's camp. . . . One time Watson got a package that smuggled in videos of his wife and kids. I asked him how he did it, and he said, 'I've got a friend on the Governor's Task Force.' . . . There's also a millionaire that owns a . . . corporation—he's made it so that Watson is 'hands off' from the guards."

"Is Watson into Satanism?"

"I don't know enough about it to say one way or another. . . . P.S. We need to talk about an affair Watson's wife had with one of the prison guards. I think that knowledge may be the main reason they put me in solitary."

Trouse's next letter came certified.

> *Dear Mrs. Tate:*
>
> *At approximately 12:15 P.M. today, I was paged to report to the Control Security Squad. The officer handed me a stack of mail I had sent to you; they were all marked "insufficient postage." Mark my words, they removed the stamps to keep my letters from you. . . . I have to tell you I'm more than a little scared of my situation here. . . . From this point*

forward, I am no longer able to indulge any information to
you via mail or phone. Anything we need to discuss must
now be done in person. . . . I sent you the visiting forms.
Please fill them out and allow a week to get them cleared.

Respectfully yours, Steve

It was time to pay a visit to the prison, but I wasn't going alone.

Sandy, a friend and press correspondent, answered my call on the first ring. "Hey, baby doll. What's shaking? And please tell me it's a good lead because I've been dry for weeks now."

"I want you to come with me to CMC to monitor an interview with one of the inmates. I'll give you the exclusive."

"Let's see, CMC equals Watson, and that must equal Steven Trouse?"

"Shit, does everyone know about this guy?"

"Well, speaking of shit, he was making a pretty big stink up there for a while. The only thing everyone doesn't know is that he must be your snitch."

I laughed. "I love that—the old bitch has got a snitch. So, will you go?"

"Wouldn't miss it. Listen, I've got to run. If I fax you my info will you put in my application for me?"

"Yep."

In an effort to conceal my rendezvous with Trouse, I filled out the traditional visiting forms for both me and Sandy and sent them off to the California Men's Colony.

Nine days later, Sandy received her letter of approval from the prison. Nothing came for me. Nor did anything come the next week. Phone calls to the prison were futile. I wanted to trust that it was an innocent mistake, yet it suspiciously rung of a counterplot to curtain my link to Trouse.

With nothing to lose but time, I made the scheduled trip with Sandy to the prison. Sandy pulled her Honda to the curb outside the administration building. With the third cigarette of the hour dangling from her lips, she swept back her long hair, turned down the Elvis Costello tape—the songs of which I now knew by heart—and pushed in the lighter. "I'm going to hang here, smoke, and read all the bylines I didn't write. Come get me when they approve your shady ass, Miss Ma Barker."

"Funny," I said, closing the door on the smoke cloud.

THE PRISON OFFICER handed me my license. "Ma'am, you need prior approval for visitation."

Graciousness was not on the tip of my tongue, still, I managed to cajole him. "Well, honey, I know that. I sent my application in over three weeks ago. They seemed to have lost it, so I thought I would come on up and try to get this all straightened out."

Slightly warming to me, he took my license back. "You can wait in reception."

The guard found me an hour later. "Mrs. Tate, they are having some time up there finding your application. I think it would be best if you came back another day instead of waiting here. This could take hours."

My smile remained, but my voice crisped, "Darlin', I'm on the Advisory Board of the California Department of Corrections. They shouldn't be having any problem approving me. Now you go back and tell Warden Estelle that I'm not leaving his facility until I've seen the inmate I came to see."

He let out an exasperated breath. "Yes, ma'am."

A diverse collection of visitors jammed the lobby. I surveyed them face by face, personality by personality, until I got the distinct feeling that I was the one being inspected. My motive for people-watching altered. Turning, I caught sight of the observer on the opposite side of the bulletproof glass that dissected the prisoners from the public. He stood fifteen feet beyond the front desk with his arms folded across his chest, his feet slightly apart, his posture unmistakable. He'd stopped with the intent of surveillance.

My training with the CDC taught me to avoid eye contact with the prisoners in passing, and never let them smell fear. I ignored the first installment and fixedly stared back. He's decidedly familiar. I searched my memory and the hint turned to recognition. His features had changed little since his last published pictures in 1969. Bobby Beausoleil is his name. Another Manson Family member in prison for murder, and he worked in the chapel. He was so brazen. Had he come to intimidate me?

Verbal communication was unnecessary since our body language served as an adequate solution. I squared off my shoulders toward him, and with just my left eyebrow arched, I scolded him with my index finger. He held his stance for a moment longer before smiling with a decisive shake of his head. His actions were as strong as if he'd actually said, "Old lady, you have no idea what you're getting yourself into." His grin expanded to laughter. He tipped an imaginary hat, gracefully bowed, and then sauntered away. In one of my less amicable moods, I flipped him off.

Considering the confrontation, Watson must have found out I was on the prison grounds and sent one of his emissaries. Remembering Trouse's last letter, I decided to take his safety, as well as my own, a bit more seriously.

The next three hours passed without incident until the guard returned. "Mrs. Tate, you've been granted visitation. Follow me please."

The officer chaperoned Sandy and me to an empty cafeteria of bolted-down aluminum tables and benches. Before I could question the odd setting, Steve Trouse arrived in the company of another guard. Shifting ahead of them, a suited man overstepped the introductions. "Mrs. Tate, I'm Larry Kamien, program director and public information officer for CMC. I'll be monitoring your visit with Mr. Trouse."

I appraised the afternoon's next stone wall. If he'd entered my hair salon, Kamien's weaseled features would have inclined me to gel his cowlick and shave off the rest of his unevenly trimmed mustache. I pointed at Sandy. "She's a press member. Under Title fifteen of the

CMC regulations, this man is entitled to a private conversation with any media person or legal representative."

"Yes," Kamien sniffed, "but you are here without prior approval. Now, would you like to have this visit or not?"

"Don't worry about him," Steve interjected. "They're all calling me a snitch already for talking to you. I've got nothing to hide. They know I'm telling the truth."

I looked back and forth between the two men. "Okay. This is Sandy. She tagged along to record everything we discuss here and as a witness in case we ever need one later."

Trouse nodded eagerly. "Good, I want everything documented. They're trying to get to my wife now. Both Watson and Davis's wives are bombarding her with visits and phone calls trying to get me to drop my appeal." He continued, hardly taking a breath. "They're running scared now. All the literature found within the chapel that carried Watson's or Davis's name has been confiscated and destroyed by prison staff. They went through all our cells and took everything. Do you still have the copies I sent to you?"

"Yeah. I'll keep them in a safe place." I looked at Kamien. "As long as you're here, I'd like to ask you some questions."

He smiled tightly. "That's why I'm here, Mrs. Tate."

I pulled out the prison codebook. "It says in here, that no prisoner shall counsel another without a supervisor in attendance. Is this correct?"

"Yes, it is," Kamien said.

"So why is Watson allowed to break that code?"

"Mrs. Tate, you have to understand that our regulations allow inmates to preach here."

I tapped the codebook. "Okay, but not in an unsupervised environment."

Kamien referenced the book, too. "Those are our regulations. If this is happening, and I'm not saying it is, the worst we would be doing is breaking our own rules. We're not apologizing for this situation, but the warden is looking into it. He's looking at everybody, not just Watson."

Trouse angrily rose. "Why don't you tell Mrs. Tate how Charles gets away with having authority over another inmate, where he can preach, teach, and discipline somebody?"

Kamien sidestepped the issue. "If you will look in General Order thirty-six, it says participation in any of the chapel or religious programs is strictly voluntary. If you don't want to listen to Charles, you don't have to stay. There are also outside people holding services over there. Go to another service."

"You're not answering my question, so I'll ask another one," Trouse affronted. "Why isn't Rev. McGuire there supervising the chapel when he's supposed to be?"

"We don't have timecards around here. To the best of my knowledge he puts in about forty hours a week," Kamien responded.

"That's a lie and you know it!"

I shortstopped the argument with a calmer approach. "What if they want to go at the time when Watson is holding the service. What if that's their only free time? I mean, these men are in need of real counseling, okay, not some jailhouse religion that's potentially feeding them a bunch of trash."

"He's not in the chapel every day," Kamien defended. "With Watson's personality, there are just a lot of people that don't like him. You don't have to like Watson. I really don't care one way or the other."

Kamien was a useless endeavor, moreover, if I didn't take him off the defensive he was liable to call an end to our meeting. I eyed Steve. "Honey, let's talk about what I really need to know; tell me about Watson and Manson."

"Okay, shoot. What do you want to know?"

"Tell me how Watson's able to manipulate people so well. How is he able to win people over?"

Trouse leaned back in his chair, stretching his legs out. "It's like this. He'll come up to you and say, 'How you doing, buddy? My name is Charles Watson.' He hates to be called Tex. Then he'll ask questions to open you up, like, 'How long you been here? How much time you got?' Then he tries to be your friend and builds you up. Telling you what a great job you're doing and so forth. He keeps building you up because there is something else he really wants off you. Then, once he has you, he starts using his influential power against you. Like when he catches you doing something wrong, or disagreeing in what he has to say. Then he attacks you by using Scriptures from the Bible to support what he's saying. I mean sure the Scriptures say that, but not in the way he uses it. Understand?"

"Only too well. I guess Manson taught him something after all. Did Watson talk to you about the murder of my daughter?"

"No. See, Charles—around me at least—would never bring that up. The only thing he ever said about the Family, and I've heard him say it many times, Charles Manson was not the brains of the organization. It was all Charles Watson; he was really the brains."

One question had been on my mind for a long time. I'd heard speculation that either Watson lied about how the murders took place or someone had gone to the house after the killers had left and changed the crime scene. "Did Watson ever tell you that Manson went back up to the house with Bruce Davis after the murders?"

For the first time since the meeting began, Trouse's confidence dropped. His leg jiggled with nervous energy. "I have no comment on that. Let's move on. Something was said, but I don't remember." His eyes scampered around, and then he whispered, "I can't give you nothing on Manson. His reach is too far."

"That little bugger? Come on Trouse, what's he going to do, put some voodoo curse on you? Don't you think if he had any of those powers I'd have pinholes all over me by now?"

"You don't understand, okay? That case is closed. Now, on Bruce Davis's part, I only remember one thing that he told me. We was out walking in the yard. He stated something like, if he ever got the chance to go back in court, he would tell the truth. And I says, 'What do you mean? I thought you did that already.' And he says, 'Specifically, I would tell the truth about where the bodies and stuff like that are.' I told him I thought all the bodies had been found, and he says, 'No, they weren't, and if I ever got a chance to get back on the stand I'm gonna tell the truth.' "

"If he's such a true Christian, why wait to go to court? Why doesn't he just call in the authorities and spill his guts?" I asked.

Trouse laughed. "Well, you and I both know that they ain't really Christians. They both got their own strategy. Like Watson, he says when he gets out he's gonna be rich and famous."

"Why is that?" I asked.

"Because of the notoriety of his case. He said to me and many others, 'I won't have to work except a little at my mom and dad's business. Then I'll be going around preaching to everybody, because all they want to hear about is Manson. So why would I have to work?' See, he's an expert manipulator of the whole system, the board, prison staff, free staff, and the inmates. I believe he collects damaging information on them all and uses it against them for his needs. All I know is he's got someone up high by the balls because his wife had an affair with one of the guards and he knows they don't want the word spread on that."

"We probably don't need to talk about that right now." I tried to squelch the story in front of Kamien.

Trouse was on his feet again. "I'm gonna open up. I could care less what they say or do. I've already been written up and thrown in the hole for 'Unlawful influence against the staff.' Funny, isn't it? Watson's the one holding this over their heads. I just bring it up in my appeal and I'm punished! I want to address the problem with you two right now while Kamien's here. Everyone knew Charles was having marital problems. So one day I saw him all upset and I asked him what was wrong. He told me, 'My wife is no longer in love with me. She wants a divorce.' When I asked him why, he says, 'She's been having an affair with one of the correctional officers and he's spending the night with her. I told him he'd better stay away, but they're still carrying on.'" Trouse snickered. "What do you think about that, Kamien? Now everybody knows. There's no more sweeping it under the rug."

"Whatever you say, Mr. Trouse," Kamien condescended.

"Mr. Kamien, has anyone investigated this?" I asked, provoked by his nonchalance.

"Of course." Kamien folded his arms across his chest. "We interviewed inmate Watson about it. He denied the allegations and felt that Mr. Trouse had a personal vendetta against him."

I leaned forward. "It seems to me that since this has become a disciplinary action against Trouse you must have spoken with others in the course of your investigation?"

"I'm not going to answer any more questions on this issue."

"Why not?" Trouse asked.

"I will not sit here and be grilled by you in front of these people, Mr. Trouse."

"Fine, I'll just tell them. Kristin just woke up to it one morning. I think she figured out Tex was using her, too, and that's what drove her off."

"Has she been away from Watson?" I asked Kamien.

"I have not noticed her absence from the institution."

"That's not true," Trouse said. "Just recently she's been gone a whole month, but she always comes back around board time because Charles is under such pressure to get out. She's back now 'cause he's due to see the board soon. Right?"

"Next month," I answered.

"Yeah, well, they put on a pretty good show for you all. That's all I can say. I'm telling you the truth, Mrs. Tate. She had the affair. You saw Kamien's response. He's scared because it's the truth."

"I'm going to call an end to this visit," Kamien said.

I gathered my belongings. "Okay. Trouse, I think the main thing here is that Watson should be pulled out of that ministry. Right?"

"I agree with you one hundred percent. That's why I started all this. I want to take it all the way to court. Can you come back sometime soon when we don't have a watchdog with us?" Trouse asked.

"Yeah, but first I'm going to Sacramento with this," I said.

I put my hand out to Kamien. "I'm sorry things got a little heated here. Before I leave I would just like to feel that Trouse will be protected."

"That's our obligation; we become liable if we put him in a situation where his life is in danger," Kamien replied.

"I'm already in danger," Trouse shot back. "They've moved me to B-quad, where I don't have any friends. Everyone thinks I put in for the move so I would be protected from Watson. They're calling me a snitch."

"Is Watson in B-quad?" I asked.

"No, he's in A," Trouse said. "I'm probably going to get a bus ride for trying to break up Tex's little cult. I've put this institution under a lot of heat by bringing you into this. You're a thorn in their side."

"Is he going to be transferred?" I asked Kamien.

"The issue is where he can safely be housed. Just as you are concerned, Mrs. Tate, we wouldn't want to do anything that would jeopardize his safety."

"I want to state in front of you two and Kamien that I am in jeopardy," Trouse said. "If anything happens to me toward bodily harm on the yard or anything like that, I hope that you will follow through on it."

Sandy finally spoke up. "Does anyone know how Watson feels about all this?"

"He wouldn't speak to me last night," Trouse answered. "He knows I'm pulling his covers. He's scared, and that makes me scared. Just yesterday, someone came down with a message that Watson wanted me to meet him in the chapel. Well, no way I'm going over there, so I don't know what that was all about, but I don't like it."

"Okay, let's go, Mr. Trouse," Kamien said.

Steve took my hand. "Don't leave me hanging in here."

"I won't. You take care, honey, and God bless."

THE TRIP TO Sacramento was unnecessary. In fact, I didn't even have to pick up the telephone. Word of my visit with Trouse swiftly traveled through the CDC, and the word must have been damaging.

Before a week's passing, I received a nondescript envelope. Enclosed were two items. A page entitled, "Statement for Release by Warden Wayne Estelle," and a note. The handwritten text lacked a salutation as well as an endorsement: "Reverend McGuire has asked to take an early retirement. Instead of twenty-five years work record, he will receive twenty-one. You have my word, Watson will be moved within the next year."

I read the press release.

"Charles Watson is being removed from his assignment in the Protestant chapel at CMC. He will receive an appropriate work assignment at Post Board Classification.

"The results of a preliminary investigation in the chapel reveal no major mechanical discrepancies. Some changes will be made to eliminate any perception by inmates or others that the chapel program has been misused. Changes will also be made to eliminate procedural shortcomings.

"Chaplain Stan McGuire has run an acclaimed program for twenty years at CMC. His dedication to filling religious needs has been reported by the press on numerous occasions over the years.

"Inmate Watson's crime has been given an accurate description by the Parole Board in past years. His term in prison should reflect no more or any less consideration than the other seven thousand prisoners at CMC."

15

BASTARDIZING THE LAW

Mr. Watson, the night that you broke into my daughter's
home you said, "I'm the devil here to do the devil's work."
As far as I'm concerned, sir, you are still in business.

—DORIS TATE

Patti

Amid a snarl of minivans, SUVS, and station wagons, parents myste-
riously lose all sense of civility in the quest of dropping off or picking
up their kids from school. Such was the case at my daughter's middle school at three in the afternoon. Horns bleeped, fender-benders
cracked, and obscenities flew. All the while, children curiously appraised whether their parent was sane.

My two-year-old son, with his shoe planted firmly in his mouth as
a pacifier, found it all very amusing from his backseat view. "Bweeee!
Mama, Bweeee!" he yelled.

I pressed down the passenger window and yelled right along with
him, "Brie! Let's go."

She ran up to the car. "Hi, Mama."

"Hi, sweetie. Where's Ally?"

"I don't know," she said, sliding into the passenger seat.

"Wait, hon. I need you to go find her."

"Aw, Mama," she groaned, "you go."

"I can't leave Bryce alone."

"Hellooo," she rolled her eyes, "I'm going to be seven next week; I think I can watch him."

"Fine," I said, too tired to argue. I pointed at her to start my spiel, "Don't open the door—"

"I got it, I got it—not for the principal, not even for the police."

I locked the car behind me, and then scrambled through the throng of children. I spotted her on the side stairs. "Ally! Come on!" I hollered across the schoolyard.

She waved me over.

"It's going to be one of those days," I mumbled, dodging the opposing flow of kids.

When I reached Ally she said, "Mama, this is my friend and her mom."

"Hi, I'm Patti."

"Hi. Suzan LaBerge. The girls were just talking about having a sleepover this weekend at our place."

I quickly appraised Suzan. A throwback to the Flower Child era. Hair long, a little wild and frizzy, bell-bottom jeans, T-shirt, and sandals. My inflamed opinion may have a biased tone, but the hippie trend is not my favorite culture. "Oh, sorry, not this weekend. I'm going to New York with my mom for the *Geraldo Show*."

"We'll make it another time," Suzan said. Then as an afterthought, "Are you going to see the show or to be on it?"

Dammit, I have a big mouth. "Well, my mom is going to be on it, I'm just going to watch." I took Ally's hand. "Let's go, hon." Then to Suzan, "I've got the kids waiting in the car. It was nice meeting you guys."

Suzan, however, was persistent. "Why is she going to be on television? Is she famous?"

I felt the tug of worriment. The invasion of privacy. "Sort of, my sister was murdered so my mother's a victims' advocate." If nothing else, the word *murder* should have quashed her interrogation just as it did with all the others who dared to ask questions.

Suzan's smile faltered. A murky sheen seemed to have glazed her eyes. "What's your sister's name?"

I gnawed at the last of my thumbnail and pulled Ally closer. "Sharon Tate."

Her eyes widened as did her smile, which I could have sworn was tinged with menace. "I knew you looked familiar; you and Sharon could have been twins. You're never going to guess who I am."

I took a protective step in front of my daughter.

"Rosemary LaBianca is my mother!" she blurted out.

"What?"

Her head bobbed enthusiastically. "As soon as you started talking about your mom, I recognized you, but I wanted to be sure."

"Oh, my goodness," I said, able to breathe again. "I was afraid you were involved with Manson." I wrote my number on her daughter's notebook. "Listen, I have to get back to my other kids, but here's my number, call me."

"Better yet, why don't you let Ally come home with us, and you can come by later for a glass of wine?"

"Well."

"Please, Mama," Ally said.

I looked back toward the car, uneasy about leaving the other kids for so long.

"Just this once. I'll be okay," Ally coaxed.

"All right. I'll come by around seven?"

Suzan ripped a sheet from a spiral notebook. "Sounds good. Here's our address."

"I'LL BE DAMNED," Mom said, when I told her about my encounter with Suzan LaBerge. "I always wondered what became of those kids—you know her and her brother, Frank, were the ones who found their parents' bodies. If I recall, Suzan had a nervous breakdown afterward. How did she seem?"

"Fine, I guess. We didn't get a chance to talk. I'll get the scoop later."

Just after dinner, the phone rang. My mother was close to hysterics. "Calm down, I can't understand you," I told her.

"Get Ally out of LaBerge's house. Now!"

"Why? What's the matter?"

"Steve Trouse called. They're friends. He overheard Watson talking to her."

"Slow down. Who was Watson talking to?"

"They're friends, Patti! Suzan and Watson are friends! They're planning—"

I dropped the phone and ran to the car. Twice, my hands missed the ignition. Finally jamming in the key, the car roared to life, and my tires squealed in protest all the way out of the driveway.

The two-mile drive to the LaBerge's felt more like twenty as I ignored red lights and stop signs through the neighborhood streets. "What in the fuck have you done?" I yelled at the rearview mirror. "Oh, Jesus Christ, you fed your daughter right to Sharon's killers!" I pounded the steering. "You broke the rules dammit. You broke the goddamned rules!"

Rule number one, the kids are never out of sight, except at school and Mother's. Rule number two, never overlook danger. Because of LaBerge's background, I'd overlooked both.

On the next turn, my tires swerved out of control along with my mind. I got the car back under control, but my thoughts continued sliding toward the cliff. What is LaBerge's background? Is she a member of the Manson Family? I never did believe in coincidences, why else would she be living so close to me. My mind ran to the worse possible places. She and Watson must have planned to kidnap Ally to silence Mom at the parole hearing.

The car bottomed out at the base of the LaBerge's steep driveway. The sight of the kids playing in the yard didn't slow the terror, only prompted tears. I'd scarcely gotten the car into Park before I was out and running. I dropped to my knees in front of Ally. "Are you all right?"

"What's the matter?" she asked.

I squeezed her. "Nothing, we just need to leave now."

Scurrying to avoid a confrontation with LaBerge, I took Ally's hand and all but dragged her to the car. Suzan came outside on her porch, smiling at first, then suspicious as she watched our frantic escape. "Patti, where are you going?"

Not until Ally was securely in the car did I turn back to Suzan. "How could you?" I slammed the car door and then left her in a cloud of dust from my spinning tires.

I went from the sanctity of tucking my kids into bed, with wishes of sweet dreams, to Mom's nightmarish history of Watson and LaBerge's relationship.

The two met through Watson's mail-order ministry. Watson corresponded with her for close to a year before LaBerge visited him and confided that she was Rosemary LaBianca's daughter. Watson had kept their blossoming friendship a secret until Ally went to the LaBerge's and a game of phone tag commenced.

Suzan called Kristin Watson with the news of our encounter. Kristin then called Tex. When Watson received the illegal call in the prison chapel, Trouse was there to overhear his joyous reaction and eavesdropped on Watson and Bruce Davis.

"What was that all about?" Davis asked.

"Bruce, my friend, I just might have two victims in my corner for my next hearing," Watson boasted.

"How do you figure?" Davis asked.

"Susan LaBerge, née LaBianca, is my ace in the hole."

"What do you mean?"

Watson ruffled Davis's hair. "Buddy, she's gonna testify at my next hearing—in my defense," Watson laughed. "And, if the good Lord Jesus is shining down on me, so will Patricia Tate. The Tate kid is playing in Suzan's front yard as we speak!"

Mom and I discussed coincidences and odds. What were the odds that out of close to five hundred cities in California, LaBerge would end up in my small community? What were the odds that our kids would become friends at school? "Has LaBerge forgiven the rest of the killers?" I asked.

"No." Mom was quiet for a moment, then said, "Now that the cat's out of the bag, I don't think they'd do anything, but keep the kids close and tell Ally to stay the hell away from that LaBerge girl."

"You don't think—"

"Shit, Patti, I don't know what to think, but I don't believe in coincidences."

Doris

In 1976 California's then chief probation officer James Rowland introduced the first victim impact statement in Fresno County. The supreme court identified three types of information to be contained in these statements: information about the victim and his life; the impact the crime has had on the survivors; and the opinion of the victim as to the appropriate sentencing.

In 1982 the Victim Impact Statement was entered into law. The statute was never intended to be used by champions of the prisoner as a ruse to undermine the Victim's Bill of Rights. Nevertheless, LaBerge planned to bastardize the law by exercising her victim's right with a plea for Watson's freedom. Surely, Rosemary and Leno LaBianca would have resisted this intolerable representation. Even so, by law, the state was powerless to intervene.

The scene was set for Watson's 1990 hearing. Within the beige cinder-block walls, a T-shaped table was almost filled to capacity. Steve Kay and I sat at the top of the cross with Suzan LaBerge behind us in the overflow chairs. On the table, in front of each of the three panel members, lay a copy of Watson's eight-inch-thick file, and between the files, Styrofoam cups that must have held bitter, lukewarm coffee, since Supt. Bob Carter winced at the taste.

I hadn't seen the presiding board member since 1982. His skin had barely aged, but his eyes couldn't outfool time. He put on his glasses to read a note that Board Member Ascido handed him from his right before he passed it to Ramirez on his left.

Behind the three men, open windows stretched along the length of the room, teasing a prisoner with the power of freedom to which they held the key. Next to a thirsty plant in the opposite corner, the video camera lightly hummed as it recorded for the press pool outside. The only thing missing was Watson.

"Charles Watson's attorney, Mr. Jan, is delayed in traffic." Ramirez's pudgy fingers crumpled the phone message and then impatiently kneaded his shiny scalp. If Jan held up the hearing much longer, he'd miss his flight to Sacramento.

A hand tapped at my shoulder. "Mrs. Tate?"

I turned to face LaBerge.

"Doris, why do you do this? Don't you think the time has come to get over this and move on with your life?"

"My dear, I will never get over this. I don't have that power. This man that you are here today to defend, slaughtered my daughter—and your mother—without an inkling of humanity and a world away from the Holy Spirit."

"But God *has* forgiven him, why can't you?" LaBerge asked.

"That is between him and God, it has nothing to do with the laws of this state. We have to have deterrents whether God has forgiven them or not, otherwise we'd have nothing but chaos. . . ." My voice trailed off as LaBerge glanced away, distracted. I closed my eyes and felt his presence behind. Dear Lord, grant me the courage to fearlessly confront the past and the demons that sealed this fate. My hands remained locked in prayer as my eyes opened in time to see Watson's reassuring wink to LaBerge. I leaned to the right to disrupt their reunion, and for just a split second, Watson caught my glare. He tugged at the lapels of his jean jacket then turned to face the board.

"Mr. Watson, I know you've been asked questions in the past," Carter began the hearing, "in fact, I've sat on your panel before and asked you a great number of questions about the murders you committed, but I'd like to know if you want to make any comment regarding your life crimes?"

"Only that it makes me sick to sit here and listen, knowing it was me that did this, and I take full responsibility for the crime, and for these people to have lost their lives so needlessly. It really hurts. But," he added nonchalantly with his lips pursed, palms upturned, "there's not much I can do but feel these feelings that I feel, and continue to go through the pain that I go through. It's tough on the heart."

"At the time of these crimes, did you form the opinion that you were going to be put to death?" Carter asked.

"No. My state of mind was that there wasn't going to be a tomorrow. That there wasn't going to be any punishment. I came to the conclusion that the end of the world was coming."

"Let me interrupt you," Carter said. "I've read the transcripts from past hearings. It's quite evident that you and your crime partners place the total blame on Charles Manson."

"I don't place the total blame on Manson, but if there hadn't been a Charles Manson, we wouldn't have done what we did."

"Well, what if we had Mr. Manson here and he said, 'I would not have committed the crime had it not been for Mr. Watson?'"

"Well, I don't truthfully think he could say that," Watson smirked. "You have to remember that at the time of the crimes, I was twenty-one and the other victims were young as well. Well, not victims, although I guess we are all victims of this case. But uh—"

I startled everyone when I smacked the table in a reflexive backlash to Watson's blunder. Carter gave me a warning look over his glasses while Watson continued, "Manson had us in a delusion. We were committed not only to kill for him, but to die for him."

"Weren't you second in command?" Carter asked.

"We were all followers. Even though I've matured and grown over the years, I didn't have leadership qualities at the time. We were deceived and manipulated by a person that was a con man."

"But it was a group action. It wasn't one man."

"Yes, it was, but—"

"All of you believed what you were talking about. All of you believed in murder," Carter stated.

"I take responsibility for that, but at the time, I would have done anything for Manson. I was willing to die for him. He had a knife to my throat—"

"Yes, we've heard that story already. Mr. Ramirez, why don't you take us into the psychiatric reports."

"Okay. Most recently there's a report that was prepared on February 23, 1990, and a prior one that the panel conducted for your last hearing in 1987—you had all the staff here for that one, Mr. Watson." Ramirez smiled. "I'll just dovetail the two together. The counsel concluded that parole considerations remain premature, and your violence potential remains an unpredictable threat to society. The report states: 'Mr. Watson is acutely aware of his unpopularity both within and outside the institution. While his removal from the chapel was fraught with accusations regarding his position and conduct, he was able to maintain a good attitude and demeanor as he swept trash and cleaned windows. He hopes to work with religious ministries in east Texas someday to show people what he has accomplished. Claims he will live his life for his wife and kids in a relative anonymity. To quote Mr. Watson: 'With my ministry work, I'll only be on regional TV, not the networks, so most of my time will be spent at home.' "

Ramirez looked up from the report. "Mr. Watson, this part really jumped out at me: 'It's been during the last three years of one-on-one therapy that he [Watson] has begun to truly experience a sense of deep remorse for the crime victims and for the family of the crime victims.' What were you doing for the eighteen years before that?"

"It's not that I haven't experienced it before, it's just that the things I have been doing over the last few years and my Christian faith has really brought it home. You have to realize at the time of my crimes, the people that we killed weren't human beings to us, we didn't see them as people with feelings, and lives, and families. We didn't see any of that. I began to see it through your eyes, for instance, because of opening myself up. For eighteen years, I didn't see the depth of it. I've always had a hard time living with it."

Ramirez removed his glasses. "You talk as if you're from another planet."

The muscles constricted in Watson's throat. "Well, I'm sorry, but I think I've come a long way as far as acknowledging what I've done in this crime. The psychiatrists think I've come a long way," he said curtly.

"To what? Being human?"

"Yes. Exactly."

Ramirez looked at the other panel members. "Do either of you have any comments."

"I think we both do," Ascido said. "Mr. Carter, why don't you go first."

"Okay. Mr. Watson, why did you have children?"

"Because my wife and I desired children," Watson replied. "When we decided to have children in the 1970s, things looked different; we were on rehabilitation instead of punishment. When I would come before the board, there wouldn't be these television cameras; there wouldn't be the district attorney, and the victims' families weren't there. It looked like in the 1970s that everyone was getting a date. Then the tide started to turn in 1982. Things looked completely different. Since then, we've often wondered why we had children."

"I'm not a mathematician, but your kids are six years, eight years, and fourteen months. That wasn't in the 1970s."

"Well," Watson stammered, "our marriage began in the 1970s."

"You were married in 1979. And if you're correct on the ages of your children, the first one was born in 1982."

"Still, in 1980 it looked good because, really, the cameras and the people here today didn't start showing up until 1982."

"You lost me," Carter said.

Watson jumped in before the question was finished. "Well I'm sorry, Mr. Carter, we seem to be having a communication problem—"

"That we are," Carter was just as quick to interrupt. "Why did you have children?"

"Well, we had children because we desired to have children."

"Are you going to answer my question?"

"Well, yes," Watson said, "you're wondering how they will handle the situation, how they don't have a dad. That they'll have grandkids before I get out. We thought about all this."

"And you went ahead and did it anyway."

"We didn't think about it in advance. Like I said, the tide hadn't completely turned when we made those decisions, but I can't say that we wouldn't have made those decisions anyway. Ah, if we had looked at what we are looking at now, we wouldn't have even gotten married."

In the twelve years that I'd been attending Susan Atkins's and Watson's parole hearings, a smile had never raised my lips, not even a smirk. At the initial hearings, I used my hands to cover my trembling lips; later, my hands concealed a possible snarl. Presently, my hands tried to keep down a satisfying, barn door–sized grin.

I didn't know why, and didn't care why this group of board members hammered Watson until he was as indefensible as a dozen eggs in the middle of a catfight. But it felt damned good to watch Tex squirm in his seat, to see his jaw muscles clench and unclench, to see his Adam's apple bob with uncertainty, and to see him lick his lips with a parched tongue. Through it all, I smoothed my grin to a ruler's edge.

Carter closed the file. "I'm finished with this portion. Go ahead Mr. Ascido.

"I'm surprised that you didn't have any comments about the counsel's report, Mr. Watson. What I found interesting was the counselor's view that you are an unpredictable threat to public safety. Because what the counselor is referring to is whether you would be able to cope outside of this institution. With that in mind, would you like to tell the panel why you haven't been able to function in any institution other than this one?"

"I haven't been to any other institutions," Watson said.

"Why?"

"Well, I came here in 1972 from death row, and I've always done real well here, and I've never had any problem or reason for transfer. I think any time you transfer from one institution to another in a high-profile case such as mine, there's some fear. And when you have a family and you do things within the community with your family, you just don't want to move. Moving my family, my kids schooling, you know, everything that we're involved in. So it's not that I couldn't cope in another institution, I feel that I could. I feel I have qualities that would work out there."

"Mr. Watson, the board recommended to the staff that you be transferred to another institution. See, when an individual has completed all the programs here, it's best for them to move on. Most prisoners are transferred upon completion of programming. You're one of the exceptions. And, when you were asked to transfer, you indicated that you didn't want to go."

"Well, I feel safe here. My family feels safe here—"

"But that doesn't have anything to do with—"

"Oh, yes it does," Watson said testily. "My family has a lot to do with my life."

"Don't you think that applies to every prisoner? So that's immaterial."

Watson toyed with his file. "I think the CDC takes consideration now for family. Even legislature with a new bill that says you have to take consideration of my family and them being a part of the community and us being closest to our family."

Ascido looked at the other two on the panel. "That's news to me."

Watson searched his file. "Really, there's a Senate Bill 2190 that requests that inmates be allowed to be in the community closest to their family within the point guideline."

"That's not the current law," Carter said.

"Yes it is," Watson insisted.

I scribbled a note and passed it to Steve, who passed it to Ascido. "The law that you're speaking of is specifically for parole violators. Now, back to the issue of prison. In light of what the counselor is saying—that you can't cope—this would be a good test. Don't you agree?"

"No," Watson shook his head, "I do not agree."

"Well, it may be the only way you get out of this particular prison, Mr. Watson, so you may want to revisit that thought. Back to you, Chairman Carter."

"It's time to hear from Mr. Kay. After him, Mrs. LaBerge, and then Mrs. Tate."

"Thank you," Steve said. "I've seen each of you at various board hearings for these killers, and I know you're familiar with the crime details; however, there is one detail on the night of the Tate murders that I'd like to share because it shows Watson's thought process.

"The incident occurred right after Watson and the others left Sharon Tate's house. Watson began looking for a place to clean off the blood. He eventually pulled up in front of the home of Rudolph Weber because there was a garden hose out front.

"At the trial, Mr. Weber testified, 'Mr. Watson was calm, cool, polite, gentlemanly, and apologetic for awakening my wife and I.' Now, this was ten minutes after murdering Sharon and the others. So, believe me, he can sit here during a parole hearing and sound like he's a reasonable person, heck ten minutes after killing five people he could sound reasonable.

"Watson is so unpredictable it's scary. There's something within him, maybe a Jekyll and Hyde personality—he can be the world's nicest guy at times, and he can be the world's most bloodthirsty killer at other times—so how do we ever know what's going to happen with him? We don't, and I think that's what's troubled the psychiatrists."

Steve pulled out a picture taken in 1969. It's a wild-eyed depiction of Watson, with long hair, his face covered with a beard, mustache, and dirt. "No matter what he looks like now, or how he portrays himself, this is what lies beneath the façade of Tex Watson. I leave it to you gentlemen to do the right thing and give him a three-year denial before his next hearing. Thank you."

"Mrs. Tate, I'm going to excuse you while we hear from Mrs. LaBerge, and then you'll be given the same courtesy for your statement," Carter said.

I hurried out the side door to the monitor in the atrium to watch the spectacle of Rosemary LaBianca's child pleading for her killer's freedom. I couldn't put my finger on it, but there was more going on here than mere forgiveness. After all, LaBerge had not contacted Krenwinkel or Van Houten, nor had she made a public plea for their release. A reporter once told me that Watson and LaBerge were friends prior to the murders. At the time, I'd thought it was speculative nonsense. Now, I gave the possibility more validity and wondered whether she or Watson actually wrote her script. LaBerge's flowered dress cherubically disguised what must lay beneath her costume; a well-concealed evil.

I was back to hiding a snarl, maybe even fangs.

"I'm grateful for the opportunity to share my feelings and point of view with you," LaBerge said. "I hope you'll understand that I put it all down in notes so that I wouldn't forget anything."

Watson watched like a proud parent as LaBerge read her pages in a monotone cadence. "I've had much time to think about the crimes Charles has committed. They affected me as much as anyone else who loved any of the victims. It has taken a lot of time, information, and God's love to come to the opinion that if this case is going to continue to be viewed in the public, they deserve to know another side of Charles's life—the loving side. They should be made aware that he is nothing like the news media portrays him. They should be aware that he is not rotting in some cell, and that he's pressing forward to become all that God has created him to be.

"I believe that Charles's positive progress should be made known and acknowledged by all. I believe he has set a precedence with the way he has chosen to conduct his life. I believe that he has made the only restitution that he can and that it is valuable for society to know that one individual made a choice to overcome his past.

"I believe fear is the primary reason Charles has not been given a parole date. Fear of recidivism, fear of setting precedence, fear of public opinion. I realize that some prisoners who are paroled return to prison within a few years, and that concerns me because I doubt their rehabilitation. I don't view Charles in this light. His life before his crimes was well rounded and rooted in good morals and beliefs. He was reared in a secure and loving environment. For only a very brief time did Charles step out of his beliefs to experience another way of life. He was lured into the drug-crazed culture of the sixties . . ."

Three pages later Carter interrupted. "Mrs. LaBerge, how many more pages do you have?"

"Just a few more."

"Okay, proceed."

"Charles made very wrong choices for his life; however, the choices he could have made after his crimes were endless and he chose to repent. I believe that this has not been an easy accomplishment for Charles, considering who he was, what he did, and where he is.

"The awareness of his change, growth, and true repentance should be obvious to you. I don't think any kind of fear is justifiable for keeping Charles in prison. For Charles, I believe twenty-one years of imprisonment, and his having to live with the memory of what he did, is punishment enough. It is my belief that Charles could live in society peacefully and should be given a parole date. Thank you for letting me participate today."

"You're excused at this time, Mrs. LaBerge," Carter said. "We'll call the other witness."

I'd been chomping at the bit for two hours—even two years—to say my piece. I took the position at the helm of the table and went at it. "I might say that I feel sorry for this man, that he chose this way of life, but there is no way of turning back, okay?"

I looked to my well-prepared statement, then chanced another plan of attack. "I'm not going to read my speech," I told the board, and then broke the rules by directly addressing Watson. "What mercy, sir, did you show my daughter when she was begging for her life? What mercy did you show my daughter when she said give me two weeks to have my baby and then you can kill me? What mercy did you show her?

"For twenty-one years, I would have liked to have asked, why? You did not know my daughter; you had never seen my daughter—at least not that I know of, she had only been home for two weeks. How can an individual, without knowing, without any abrasive feelings go in and slice them up?

"Sharon was eight months pregnant. What about her family? What about her family? And what about the family she was going to have, sir? Is that not a family? I saw how your feathers ruffled when you were talking about your family. What about *my* family? When will Sharon come up for parole? When will I come up for parole? Can you tell me that? Are these seven victims, and possibly more, going to walk out of their graves when you get paroled?"

I took a calming breath. "Yes, I feel sorry for anyone that has chosen that way of life. But if you can tell me one person that wants to live next door to you, with your children, and with their children, I'll eat my hat. You cannot be trusted.

"And I'd like to remind the board that religious faith has nothing to do with release. If it did, we can open the doors for all inmates throughout California. Let's open the door for Ramirez, let's open the door for Randi Craft—well, we can't do that because they are on death row. But let's remember back when this man was on death row, before a fluke when the death penalty was overturned, and his death sentence was commuted to life. It took California only three months to get enough signatures to get that death penalty back on the ballot. I do not believe that the people of this state could accept the fact that this serial killer could be released to society. I rest my case."

With that, the parole board adjourned to decide Watson's fate.

They deliberated for twenty-five minutes before unanimously denying Watson parole. He would not be eligible again until 1993.

Carter slid a copy of their decision to Watson. "The prisoner needs to show he can cope under any circumstances, in whatever institution he's assigned. When safety is not an issue, the prisoner is to cooperate with a transfer to another institution and the staff is being requested to transfer the prisoner to another institution for further programming as soon as it is feasibly possible. The prisoner is released."

When Watson stood to leave, his legs faltered and he dropped back into his chair. As he rose again, our eyes met. He looked as if he'd seen a ghost, and he had. The ghost from his past came to haunt him during this parole hearing. It was going to be a long time before Watson felt peace of mind. Welcome to my world.

Suzan scurried out behind Watson.

Mr. Jan shook my hand. "Mrs. Tate, you're my hero, don't you ever forget it."

"I'm no hero, honey. I really don't understand how you can represent him.

Jan shrugged his shoulders. "Everyone is entitled to an attorney, Mrs. Tate, even killers like Watson. I'm just doing my job."

"It'd make all our lives a lot easier if you all weren't so damned good at it."

At the parking lot, Steve and I ran into Suzan LaBerge. He blocked her path of escape. "You know, Suzan, that you dishonored your mother today."

"No I didn't."

I couldn't hold my tongue a second longer. "Oh yes you did. You were there today defending the man that stabbed your mother in the back forty-two times, that's beyond human forgiveness. Every mother within the sound of my voice would cringe if their kid went into a parole hearing to beg for their killer's release. You make me so sick I can't even stand to look at you, you dumb shit."

"You have no right to talk to me that way," she hissed. "It would be wonderful if you could be like my parents."

I got in the last lick right before the press came into earshot. "Oh, go to hell."

"Mrs. Tate," a reporter approached, "Watson has been in prison now for twenty-one years, that's longer than most of the convicted killers in California. When will you feel he's served enough time?"

"Oh come on, you tell me," I lashed out, still fuming over LaBerge. "Watson has been in jail for twenty-one years, okay. Divide that by the seven people that he killed and that's three years for each person. Was each of their lives only worth three years? I'm sixty-six years old. I won't be alive forever, but I will go to my grave intent that justice be served, and future generations will not forget Tex Watson's evil deeds."

16

PASSING THE TORCH

They received the death penalty and it should have been fulfilled.
Then, we wouldn't have to worry about any of this, would we?

—DORIS TATE

Patti 1992

An errant storm front edged its way inland from the coastline with low-slung clouds brooding well below my bird's-eye view from the Holy Cross Cemetery. The June Gloom, as the Angelinos like to call it, had crossed into July. Mom wouldn't care now, but for the living we decided the only proper place to bury her remains was next to Sharon's in the same plot.

The wind kicked up. I tilted my face skyward and found the silvery clouds with heavy black underbellies hurrying across my path at time-lapse speed. My eyes stung from the snapping current that kindled more tears. Memories of the past clawed at and pierced my durable resistance while I stood over Sharon's freshly burrowed grave; her once pristinely silver coffin now blushed to the hue of the earth.

I was eleven years old again, wanting to open her casket, needing to see my eternally beautiful twenty-six-year-old sister, wishing I could lay at peace with her.

Time heals all wounds? Certainly not. The searing recoil of Sharon's murder burned through layers and years until it reached my mother's soul in the form of cancer.

"IT'S GLIOBLASTOMA," THE doctor said.

"Geo what?" Dad barked at the neurosurgeon.

"*Glio*blastoma. It's a malignancy of the glial cells that are the supportive tissue surrounding the neurons—"

"Son, do you have the capability of speaking English? What in the hell's the matter with my wife?"

"Mr. Tate, calm down," the doctor suggested.

"You try sitting in my place and then tell me to calm down."

"Doctor," I glared at Dad, "what's the simple version of your diagnosis?"

"Your mother has an inoperable, malignant brain tumor. It's a grade four, which means it's extremely aggressive, reproduces quickly, and invades the surrounding healthy tissue. Prognosis: She has ten to twelve months with radiation therapy, six months without treatment. You'll need to find a neuro-oncologist and a damned good radiologist as soon as possible." The surgeon stood, his extended hand ignored. "You have my deepest sympathy."

The tumor would spread through the adjacent brain tissue and eventually, if Mom survived long enough, throughout her central nervous system. The parade of physicians she saw refused the faintest shadow of hope. More readily available was a list of symptoms to track: seizures, memory loss, changes in behavior, progressive weakness, loss of speech and vision, fluctuating mental status, coma.

I ignored Mom's future. Instead, I chased the past, trying to recapture squandered time, and in doing so, I accused, prosecuted, and mentally condemned the smugglers of Mom's attention. I resented every victim who had called her for solace or advice, every organization that had needed her support, and every bit of legislation that had distracted her focus from me over the years. Mayhem holds no boundaries, including holidays. The phone rang off the hook with victims on Thanksgiving, Christmas, and Mother's Day while I was pushed to the back burner, stewing over those who stole her away.

They were all time thieves, including President George H. W. Bush, when he flew into Los Angeles to appoint Mom his 738th Thousand Points of Light. In lieu of viewing it as the proudest moment in Mom's life, I regarded it as an intrusion of our last weeks together.

Her motor skills degenerated and her body weakened by the radiation treatments, I pushed my wheelchair-bound mother through the airport's special security to meet the president's plane.

The unrestrained wind gusted across the tarmac while we waited for Bush's arrival. Mom shivered. The scarf and straw hat veiling her balding head were hardly adequate against the chilling dampness. I removed the hat, and wrapped my arms around her head to shield her from the flaring blasts. I was annoyed at the president for making her uncomfortable. My one-track mind concentrated on thoughts of getting her back to the comfort of her bed as quickly as possible. Sensing this, Mom tickled me under the arm. Her words came thickly. "I'm okay." Then she handed me a pamphlet. "Will you read this to me?"

Had she forgotten the reason we were here? The paper quoted Bush's passionate idea behind the award he was about to present.

"America has always been led by example. So, who among us will set the example? Which of our citizens will lead us in this next American Century? Everyone who steps forward today to get one addict off drugs, to convince one troubled teen not to give up on life. To comfort one AIDS patient. To help one hungry child. We have within our reach the promise of a renewed America. We can find meaning and reward by serving some higher purpose than ourselves, a shining purpose. The illumination of a thousand points of light is expressed by all who know the irresistible force of a child's hand, of a friend who stands by you, a volunteer's generous gesture, an idea that's simply right.

"Points of light are the soul of America. They are ordinary people who reach beyond themselves to the lives of those in need. Bringing hope, opportunity, care, and friendship."

My lids tightened around guilty tears. This substance of Doris Tate was a stranger. She'd tried to share her work with me, but I didn't want to hear it, couldn't bear to listen, didn't want to waste more of *our* time on *them*. There'd been so many sad stories coupled to a never-ending line of victims needing her support. No matter how long that line grew, she refused to turn anyone away, and therein laid my resentment—time thieves.

Mom and I rehashed the issue in clashing arguments until the impasse of my thirty-first birthday. The phone rang just as we sat down for dinner. I cringed. She came back a moment later putting on her raincoat. "That was Helen. The police think they've found her daughter's body, and they've asked her to go in for an ID."

My fork clanged to the plate. I couldn't look at her. "Mom, it's my birthday. Let someone else go with her."

"She asked me to go with her."

"I'm asking you to stay."

"What would you have me do, Pat? Call her back and say, 'Gee, Helen, I'm lucky enough to have my family safely around the supper table, so you go alone to pick out an identifiable piece of your child's mutilated body—maybe a mole that withstood her otherwise bashed-in face!' "

"Stop it! Just stop it! Your family is *here*."

"Lord knows I didn't raise you to be this selfish, Patti. I'll be back in an hour."

"I won't be here."

"Suit yourself."

We didn't talk for three weeks.

The president's plane taxied closer. I'd become my own worst enemy. Everyone except me had recognized the importance of Mom's work, including the leader of our country. I took her hand. "How could I have been so blind?"

"Nonsense. You have a right to own those feelings."

"I'm proud of you," I persisted, bathing my conscience.

She made a sweeping motion with her one functioning arm.

"Clean slate." She pulled my head close to her lips, her words flowed smoothly as if my confession had regressed the tumor. "When tomorrow comes, and it surely will for both of us, I want you to remember this—cause I sure as hell won't," she smiled, and swatted my behind.

The week following Mom's shining day with President Bush, she flipped off the Phil Donahue show. Her hand patted the bed for me to join her. The same hand softly closed around mine while she studied my face. Finally, she asked, "What are we going to do?"

The question whipped me back to August 1969. In nearly identical positions, she'd asked, "What are we going to do without Sharon?"

I didn't answer the question then because making plans without Sharon seemed like killing her in my own right. Now, I felt that continuing this conversation would be tantamount to killing Mom.

Aware of my fears, she continued, answering the question that I dared not ask. "I can't go until I know that someone will continue my work."

"Mom, I'm not strong enough—"

She forced my look to her eyes. "Stop hiding from your fears, Patti. Use them to find your strength again because pretty soon I won't be here to be your buffer."

The saddening finality of her wisdom and frightening change it was sure to cause in my life cinched my throat. Though I was thirty-five, I curled up next to Mom like a child, burying my face against her neck; deeply inhaling her comforting scent. Silent moments passed before her whisper broke through, "Darlin', if that son of a bitch or his followers are ever released from prison, justice will be lost."

Hidden from her sight, my tears rolled freely. I nodded against her neck, "Okay, Mama. I'll do it."

Two days later, Mom fell into a coma. I lay at her side every day, holding her, massaging her, rubbing her head, anything to comfort her as she'd done for me throughout my life. She appeared younger; the worry lines that had creased her forehead, eyes, and lips were smoothed by her slumber. Wherever her mind had taken her, she was at peace.

My father had changed, too. Two decades after he had moved out of Mom's bedroom, he moved back in. Camped out on the floor in a sleeping bag next to her bed, he watched over her around the clock. We didn't really need a nurse because Dad doted over Mom as I never thought he could. Brushing her hair, rubbing lotion over her dry skin, changing her bed linens and jammies, and, before the coma, spoon-feeding her the only thing she'd eat, strawberries and cream. For the first time in my life, I saw the man that Mom had fallen in love with so many years ago. He was a stream's ebb and flow of sensitivity, yet he possessed a self-protective indifference. It was as if he was a parallel current of strength and weakness.

I spent each waking hour waiting for my mother to take her last breath, yet she lingered for a week without change while I soaked up every second she remained. Afraid to close my eyes, I rebuked sleep as another time thief. The daylight hours passed to onyx, one after another after another. Over and over, I whispered to her, "You've waited so long to see Sharon, why are you hanging on?"

On day eight, depleted of sleep, I closed my eyes.—*Just for a minute*, I reasoned.

Entrancing shadows from the television danced across my eyelids. Smoky, abstract shapes stretched apart, then fingered back together. Their ballet propelling me forward until a white flash absorbed me into the past, where I find myself on the bench in Robert Evans's greenhouse. The mystery man from Sharon's funeral is again at my side, repeating, "All you have to do is call her name and she'll be there. Talk to her every day because she will need you as much as you need her."

I'm launched into Jay Sebring's house, looking for ghosts with Sharon. She stops abruptly behind a bar full of wines and liquors. Her eyes are wide and she puts a finger to her lips. "Shhhhh. I will show you the secret." She continues in a German accent, "But! You must never tell anyone or I vill have to keel you. Versthen?" She presses a secreted button below the bar's surface. A hidden door swings open.

"Wow!"

"Come! Inside is the key to the kingdom," she says, pulling me through the door.

I expect to find a treasure chest full of jewels. "Ah, it's just file boxes and paintings."

"To your eye," she says, still playing along, "but this painting of a flower holds very magical powers. Close your eyes and make a wish— keep them closed until I say you can open them."

I think of a wish while I hear her scurrying around. I feel something wet against my cheek. "Okay. Open them!"

"Oh my God, it's a puppy! Is it for me?"

"Yep, happy birthday."

Fragments of Sharon's life splash through my mind as if the key to the kingdom has unlocked my secluded memories of her—memories I had neatly tucked away with the pain of her death.

Sharon flips through *Movieland* magazine. "There," her finger lands on a small picture, "that's Roman."

"That's your new boyfriend?" I ask.

"Uh-huh."

"Is he as nice as Jay?"

"Well, he's a bit like Dad. He smokes cigars. He's very sensitive and stubborn. He makes decisions and nothing changes them—that's for sure. He's an interesting little guy, except he's short and ugly," she giggles. "I don't know if that means anything or not."

"Only if he's mean."

"He's not mean, he's marvelous."

Whiteness again, and then I'm in New York, where snow blankets the ground near the Hudson River. The water beyond us is a motionless slate of ice. The air is just as icy. "Cut!" the director yells.

From just off camera, I run to Sharon, and throw her mink coat around her, full of worry. "Why can't you wear this during the scene? It's freezing out here."

"In the story," her teeth chatter, "I've hocked my mink coat just before this scene, so all my character has left to wear is this raincoat."

Mark Robson, the director of *Valley of the Dolls*, towers over us in his winter parka, mittens, and knit hat. "My dear, the scene won't work if you're convulsing with cold."

"What do you expect?" Sharon challenges. "It's so cold I can hardly open my mouth, and when I do the wind is blowing so hard that it freeze-dries my teeth, then my upper lip sticks to them. And in case you haven't noticed, there's a blizzard *and* you've had me out here for an hour in stockings and a trench coat." She blows out a purposeful breath. "Do you see that? It's so cold that you can't even see the warm air from my lungs. It just disintegrates. So yes, I'm shaking. And yes, I understand it doesn't work for the scene. And *no*, I'm not an idiot."

"Makeup!" Robson calls out, completely unfazed by Sharon's meltdown. "Give her some color. She looks blue. Let's go again!" he yells to the crew.

Blurry at first, our kitchen in Palos Verdes comes into focus. "We'll be back in California next week." I pull the phone from Mom's ear so I can hear Sharon's voice from England. "Roman is making a film out of a book called *Rosemary's Baby*. It's a fantastic book by Ira Levin. Get it if you can, I think it's just getting to the US market. Anyway, he has a two-picture deal for directing, and he's writing the script for both—though you mustn't tell anyone. No one knows yet. Listen, I've got to run. Give my love to the girls."

I grab the phone from Mom. "I'm here, Sharon. I love you, too."

"Love you more, Patty Cakes."

A gentle hand nudges me. It's Mom's nurse, Frannie. "Honey, I think she's gone. I'm going to get your father."

I roll over; confused for a moment, and then I know I'm in my mother's bed. Her room is as peaceful as a gentle spring rain.

Like the last burning glow of a candlewick after the flame is extinguished, Mom is barely with me. I lay across her unstirring body, holding on, selfishly wishing her back.

I close my eyes. Ideas form without words, without visions, just knowledge. Her soul is leaving the body. Weight is coming off her as if she's taking off a horrible lead jacket she's worn during her lifetime. Light as a feather she rises. I feel a rush go through me, while euphoria overcomes sadness. She's gone.

Another gentle nudge. Now awake, I rolled away from my mother to find Frannie looking down at me. "Patti, wake up. She's gone. Your mama's gone to see the Lord."

While Dad stayed in the kitchen waiting for the mortuary, I picked up Mom's Bible from her nightstand. A page was bookmarked with my children's picture along with Father O'Reilly's eulogy for Sharon's funeral. The marked passage was from the book of Judith: "And they cried to the Lord the God of Israel with accord that their children might not be made a prey, and their wives carried off, and their cities destroyed. . . . Remember Moses the servant of the Lord overcame Amalec that trusted in his own strength, and in his power, and in his army, and in his shields, and in his chariots, and in his horsemen, not by fighting with the sword, but by holy prayers: So shall all the enemies of Israel be, if you persevere in this work which you have begun."

IT TOOK THE death of my mother for me to figure out her life. I let the fine grains of dirt around Sharon's grave sift through my fingers. An area at the foot of her coffin had been cleared for Mom's urn.

If President Bush's Point of Light could serve as Mom's epitaph, then the thoughts Father O'Reilly shared during Sharon's funeral seemed to foreshadow Mom's life.

"There is one question I must ask this morning because I believe that we must do more than mourn the passing of Sharon. Her talent, the memory of her friendship, and her love call us to transcend our present sorrow. They demand that we engage in purposeful action to wrest some meaning from a senseless deed. We can do something, I believe, and we must. We are the only ones who can answer for her before God and man.

"What must we, the living, do to ensure that such a terrible thing will never happen again? What must we do to bring about a world where there will be no more hate, no more cruelty, no more awful tragedy? It is left to us to determine what good or evil will come to this world. We create in every act of good we do; we destroy in every act of evil we perform. We can make the world a better place. We hold the future in our hands. In God's name, let us put out our hands to the task and our talents to the cause of the right and the good."

I left Mom's public memorial service early so that I could have a quiet moment with Sharon before the rest of the family arrived for the private ceremony. The last time I stood near Sharon's coffin, I felt a tremendous need to say good-bye. Now I was equally needy to say hello.

In my hand was an aged, worn letter, soft as fine fabric. The words from Sharon to Mom are faded though legible:

> *Dear Mother,*
>
> *Right now, I'm waiting. Roman had to go to New York for a few days, and will be back on Friday. As far as our situation, I'm waiting because I really know he wants to be a faithful husband, but he's frightened. Believe me, he's changed enough in the past year, so I'm going to give it until the Academy Awards to see how things are. Believe me, I'm just as obstinate and opinionated as he is and maybe even more so. I know this is the only man who could ever keep me interested forever. So I'm not going to give up so easily. I think the mistake with him would be to give up. You see, that would only prove to him that love wasn't so strong. I want him to understand that love is a bond that keeps us all together forever. I must say, I may appear not to be a very strong girl, but I'm not a giver-upper and what Lola wants, Lola gets, so the song goes. Well, I'd better close now. I love you all very much and write soon.*
>
> *Love, Sharon*

In my other hand, I held a letter that I wrote to Sharon last night.

Dear Sharon,

Life slips by, doesn't it. Holding on to it is like trying to keep water within cupped hands. Tonight, I feel lost. I've lost you, Mom, and myself. I always tell my kids if they get lost, stay put, hug a tree, and wait for me to find them. Here I am lost without a tree in sight—and even if I found one, who's going to rescue me? At thirty-five, I've found that I haven't any real friends, just many acquaintances, including my husband. He's the one person I should be able to open up to, and yet my life and your death are a taboo scandal he refuses to acknowledge. I guess my marriage is just one more drop that slipped through my fingers. We lost respect for one another years ago, and without respect, there's no hope for love or even friendship.

So instead of a tree, I decided to find you, and the promise I made to you all those years ago in that dressing room at Saks Fifth Avenue because maybe you can help me to find myself. That portrait of you that's been hidden in my closet by a lot of clothes and a lot of years, comes out tonight.

Love you tons and tons—and write soon—Patti

I dropped both letters onto the surface of Sharon's casket. "You take good care of Mom."

THE TORCH HAD been passed, and the gentle, circling winds of change that started with Mom's cancer zeroed in and knocked me on my ass. A part of my life had raced ahead while the other half had stalled in my oblivious existence. I was haunted by two warnings: the mission skulls on the brevity of life, and Mom's to stop hiding behind my fears. I found myself an orphan seeking identity; clawing my way back to myself.

The night following my mother's funeral, I took my first timid steps toward a new and more daring lifestyle. Just as I promised Sharon, I removed her picture from the closet and hung it prominently in the living room. Satisfied, I waited on the couch, knowing my husband would come out to see what I'd been hammering. I was looking for a fight, and he was about to unwittingly step into the ring.

When he entered, his eyes followed my gaze to the picture. "What's that?"

"That is my sister Sharon."

"Obviously. What's the deal?" he asked.

"The deal is, that's my sister and I'm tired," I said.

"Yeah, I know, we're all tired; it's been a long day—"

"No!" I cut him off. "I'm tired of everything, tired of hiding my family and Sharon. I've been hiding for so long that lately, when I look in the mirror, I don't recognize who I've become.

"I've been a shitty role model for my kids. What have I taught them? To hide their heads in the sand and let someone else worry about the bad stuff. No more. I want to be me again—that's a hoot isn't it, I don't even know who I am. But I'm going to find out whether you like it or not. I buried my mother yesterday, but I will no longer bury my family or what we stand for."

Caught off guard he said, "I'm not asking you to."

Angered at his denial, I stood to confront him. "You ask me to every time you deny who I am." Then I hit him with a verbal one-two punch: "I'm going public. Next week the CCPOA is opening the doors of the Doris Tate Crime Victims Bureau, and they've asked me to be a board member."

"What did you tell them?"

"I promised Mom that I'd carry on her work."

"So what does that mean? You're going to ignore your family like she did?"

Six months ago, I would have agreed with him. With my mood, he was lucky to be beyond my arm's reach. "Ignore us? Everything she did stemmed from love in an effort to make the world a better place for *us*; she didn't care about herself. I will honor the work she began, and if my kids have a few less minutes with me, then so be it, because in the end, it's all about them. I'm going to bed. And I want to be alone."

I tossed and turned into the dead of the night thinking about what direction I was heading. At two in the morning, I gave up and switched on the light. I picked up my new journal and wrote.

> *Dear Sharon,*
>
> *I talk a good talk, but I'm sacred shitless. I've been trying to hold on to my marriage for the kids' sake, but I can't keep up the charade. Tonight, I told him I wanted to be alone. The only thing that accomplished was equaling his physical distance to his mental distance. Here's the funny part. That situation pales in comparison to having to go and face one of those monsters at a parole hearing. I'll tell you, it's like Freddy Kruger's knocking on my dream door and saying, "Surprise, baby, this is the real deal!" I don't know how Mom did it for all those years—not just did it, but learned to relish giving them a piece of her mind. To tell you the truth, I don't think I have it in me to carry it through. Mom's probably pitching a fit about now, but tell her to rest assured, I'm not a giver-upper either. More soon.*
>
> *Love you tons, Patti*

I flopped around on the bed like a hooked fish; my fragmented thoughts skidding the surface side by side. "She's in a better place," scrambled in my ears, the same token the well-wishers had uttered since Mom's passing. Of course she's in a better place, but why did Mom have to get the cancer in the first place? Why did Sharon have to be murdered? In the end, who had it worse? Sharon's fright for an hour, or Mom's staggering disease? Why was I still here? Why was I deemed fit to go on when they weren't? In contrast, maybe death is a reward for the good while life is punishment for the bad. What direction would our lives have gone in if Sharon had not been murdered? Would Mom have gotten cancer? Would I be lying in this bed? What bed will I be in next week? How in the world am I going to support my kids?

I'm not sure when I finally fell to sleep, but I must have because when I rolled over, I found Mom and Sharon at the foot of my bed. Mom was again robust and healthy, Sharon, young and gorgeous. They appeared so very close, yet when I reached out to them, they seemed a hundred miles away.

Their appearance wasn't startling. To the contrary, it was natural and comforting. My darkened room gradually brightened from the radiance discharging around them. Their light gravitated around me, embracing me with a love beyond human emotion.

Their arms outstretched. At the same moment a silver cord uncoiled from my belly and then drifted toward them to form a connection. My body was weightless, floating from the bed, gliding across the cord's path to Mom and Sharon. I sailed further and further until the brilliance of the light erased time and space. I wanted to pass through to the other side of this light, where Mom and Sharon hovered, but as I got closer, Mom put her hand up to stop me. "Soon, honey, but not yet. For now, you've got to go back and finish what you've started."

Arms seemed to be engulfing me from every direction, but something felt different. I feared that opening my eyes would diminish the surrounding serenity. There was a fluffy kiss on my cheek. Then a familiar giggle brought me into full consciousness. Still, I didn't open my eyes; I knew the touch of my children. My son was resting on my belly giving me little butterfly kisses all over my face; the two girls snuggled up on either side of me. I pulled them in closer, tightly hugging them. I could have stayed that way forever. My children were my lifeline, my joy, and my best reason to get out of bed, put one foot in front of the other, and get on with it.

The pink and purple sky turned blue as the sun rose higher with the promise of a new day.

Bryce pushed at my nightshirt and blew raspberries on my tummy. "Mama, what's that? Is that blood?"

I lifted my head and simultaneously rubbed at my sore belly button; it was cranberry red and protruding just as it did in the days of my pregnancies. "I think it's Nana's calling card."

17

THE CROSSROADS

These people were so cruel. My sister was the last one to die. She
sat and watched every one of her friends die before her in a very, very
cruel way. And I relive her last moments in life over and over and over.

—PATTI TATE

Patti

The morning dew had yet to evaporate when the doorbell's ring disturbed my morning cup of coffee. Something only the foolish or unfamiliar dared because it was the only unclaimed part of my day.

After the inevitable breakup with my husband, I moved with the
kids into my parents' house. In the absence of Mom's warmth, the
house had cooled to a place of indifference, where Dad aimlessly
roamed from room to room like a lonesome dog awaiting its loving
master's return. Almost a year had passed since Mom died, and Dad
had not moved so much as a stitch of clothing from her closet. He'd
never admit it aloud, but I knew he missed her terribly.

Living with Dad again was enough to roust the dead from the
graveyard of my childhood. With just a look, he could whisk me back
in time until my kids weren't the only children in the house. My circumstances weren't the best; no one likes yelping defeat on the hobble

back home. Regardless, I had nowhere else to turn. Rent was free, but I paid for everything else. A friend helped me get work as an extra on a television show. And even though I had zero interest in working in the film industry, a hundred bucks a day entitled me to the best part-time job in town.

Wondering who dared the intrusion, I wiped the moisture from the window to find the postman. He waved an envelope with the green sticker of certified mail. Good news never came when a signature was required. Probably a creditor looking for money I didn't have. I opened the door with a self-defeating smile, until I noticed the return address. He instantly became the enemy. I'd only signed a *P* before pulling back the pen. Was it too late to stop? Could I slam the door and act as if he didn't exist? I contemplated my next move. His eyebrows rose as if to say, "Sometime today, lady."

My gaze turned to a glaring punch.

He shrugged his shoulders. "Listen, I just deliver it. You going to sign or what?"

Hesitantly, I put the pen back to the paper while each letter quickened my heart until it felt as if a tribe of conga players had gathered in my chest.

Nightly, since Mom had passed away, came my ritual prayer that this day would never come. Despite my persistence, destiny had arrived, but I wasn't ready and probably never would be.

There was no reason to open the letter from the Board of Prison Terms. It didn't matter which of Sharon's killers had the upcoming parole hearing; they were collectively and individually petrifying. I tossed it aside, hoping that by some miracle it would disappear.

Twelve hours later, and a lot of contemplative denial, the letter spied on me from the counter. With the kids in bed, the house was again peaceful. I ripped open the envelope. "Notice of Parole Hearing: The named prisoner will appear before the Board of Prison Terms for a hearing on January 20, 1993. Atkins, Susan, Murder First (seven Counts)."

A little more than a month away.

All day I had walked an emotional tightrope. At last, I fell off. I closed myself in the family room, where I cried, ranted, kicked furniture, and threw everything unbreakable within my reach. Panic, anger, sadness, determination, back to panic, the turmoil rebounded off the walls.

The door creaked open. It was Dad. "Everything all right in here?"

I threw the notification. "Why do we have to go through this shit?"

"I don't know, baby, it's just the cards we've been dealt. All I do know is that I can't go again."

"Why me? Why did Mom have to ask me? I'm not strong enough to do this, Dad."

"Then don't go, Sugar. Your mama would understand. I don't like seeing you like this, and I know she wouldn't either. It's not worth it. Let them walk, someone will kill them and then this will all be over." He closed the door, leaving me alone with my demons.

Mom and Sharon's lifetimes spanned the confining walls. Mom's achievements in victims' rights filled one. Another wall captured Sharon at various stages of her life: a toddler taking her first steps, the high school prom queen, stills from her movies, and pregnant with a baby she'd expected to nurture into adulthood. I eluded their reach by turning to the bookcase.

Perched on its two-decade seat, *Helter Skelter* upstaged the others with its ominous, red letters beckoning my attention. Written by the prosecutor, the core of the book held the motive and details of Sharon's murder. Before tonight, there had never been a reason to pick it up. I wiped dust from its binding and took it from its tomb. It felt as heavy as the gloomy weight upon my shoulders. Inside the front cover there was an inscription.

> *To Paul Tate*
> *There's nothing I can say to express all my outrage over what happened to your lovely and exquisite Sharon. You have my very deepest and heartfelt condolences.*
> *I have to believe that if there is any ultimate justice in*

the universe, Sharon's killers will pay for what they did
much, much more than they already have, and you, Sharon,
and Doris, with the rest of your family, will be reunited in
heaven.

Paul, I have the greatest respect for you and the moral
strength you displayed in not bringing about immediate jus-
tice in the courtroom, for which no one would have blamed
you if you did.

My warmest and best personal wishes in all the years
ahead to a wonderful human being.

Respectfully, Vince Bugliosi

What ultimate justice would there be if Sharon's killers roamed free? All of them claimed to be rehabilitated. Was this possible? At the time they received their death sentences, only four other women in California had had this penalty imposed on them, none as young as Atkins, Krenwinkel, and Van Houten. Even so, their jury was unable to find one mitigating circumstance to support even a remote possibility of rehabilitation, urging their consciences to opt for the death penalty. Those twelve jurors had spent more than nine months, eight hours a day, watching these killers, observing their every idiosyncrasy while listening to testimony. The wisdom of that jury's decision should be upheld to the fullest that our laws currently allow, life imprisonment. How parole ever became an option is beyond reason.

In 1971 Alvin Lee composed my life formula: "I'd love to change the world, but I don't know what to do. So I leave it up to you."

I had hit so many crossroads that there were none left to take. At each preceding one, I'd chosen the trail that I assumed would eclipse the past. Bar none, I found myself in a vortex that plunked me right into yesterday's tar pit, where I squirmed and tugged for release from the swallowing black goo.

I collapsed in Mom's favorite recliner, drained, defeated, and out of options; cornered into taking a leap of faith that the driving force of the past was the only way to advance.

I knew little of Sharon's murder, even less about Susan Atkins.

Mom had sat in this very chair and challenged a reporter who believed Atkins was rehabilitated. "Honey, the tears are fake," she told him. "Atkins says she feels, but she doesn't. I have not heard this woman say, 'I am truly sorry for what I did.' And if you can get those words out of her, I'll give you a hundred bucks."

"You're on," the reporter took the bait. "I have an interview with her next week. I'll send you the results—have your checkbook ready."

A month later, a video arrived from Australia's *60 Minutes* with a short note: "I lost."

The journalist had put up a good fight. "Could you say you're sorry to Mrs. Tate?" he asked Atkins.

"I don't know that Mrs. Tate could ever forgive me. My hope is that someday she will," Atkins replied.

With a hundred dollars at stake the reporter pressed, "Could you get out the words and say you're sorry to Sharon Tate's parents for what you did to them?"

Atkins took a beat. An almost imperceptible smile creased, "You ask hard questions." She paused again. "There are no words to describe what I feel—'I'm sorry,' 'please forgive me'—those words are so overused and inadequate for what I feel."

No doubt those sentiments would have been a good place to start. We all needed one. Mine was for another day. I turned out the Christmas tree lights and went to bed.

PART OF MY spirit died when I read the details of Sharon, Jay, Woytek, Gibbie's, and young Steve Parent's murder. I cloaked my activities from Dad. It was pointless for both of us to languish in a regression to 1969.

A thief in the night, I spent the greater part of a week snooping through Dad's army footlockers in the garage. His 1969 investigation was meticulous, but upon its completion, he stored it in a jumbled index, where it would remain buried in his footlockers. The dim fluorescent buzzed overhead as I scavenged and pulled the appropriate files: *Tate Progress Reports, Coroner, Witness Statements, Crime Scene, Grand Jury,* and *Media.* I stowed the materials needed for study in my closet; three boxes all told, including Atkins's parole hearing transcripts, and *Helter Skelter.*

On a day when the kids were in school and Dad was on a plane to visit relatives in Texas, I settled into Mom's easy chair with Bugliosi's book and began my odyssey. The itinerary was straightforward, nonstop to perdition.

I gazed at the doorway and relived the moment that clung to me like nurtured creeping ivy, with leaves that webbed around until they smothered me. Sometimes I managed to rip a vine free for breath. Other times, it was an unbreakable mesh that showed my mother wilting like a dying tulip as she says, "Sharon's been murdered."

My first cigarette in more than a year rattled within my fingers' clasp. I lazily blew the smoke from my lungs, luxuriating in the rewarding light-headedness that coaxed me to open the book.

I knew only the faintest detail of my sister's murder: she died from sixteen stab wounds. The rest fell on deaf ears that diligently blocked the intricacies of that mayhem. I ignored every book ever written about the topic. When the news reports elaborated on specifics, I changed the channel. If anyone talked about it in my presence, I left the room.

What I did adhere to was Mom's belief that each of Sharon's killers were responsible for their actions; therefore Manson's philosophy, how he formed the Family, or conjecture of brainwashing was irrelevant and hardly worth investigating. "Just the facts, ma'am," as Joe Friday said.

Three snubbed filters crowded the ashtray before I dared to open the book. Above the first line of text, I found a penned note from my mother:

I figured you'd start here. Remember our friend Saint
Francis's guidance: "Start by doing what's necessary, then do
what's possible, and suddenly you are doing the impossible."
I'm right over your shoulder.

Loving you always, Mother

I couldn't help but look. Of course, no one was there. Nevertheless, I didn't feel as isolated as I weeded through the pages of *Helter Skelter* searching for and then assembling the puzzle pieces that shaped a repulsive canvas.

August 9, 1969

Sharon's house on Cielo Drive is stashed in the hillside of Benedict Canyon, above Beverly Hills. Built on a two-acre ledge in the thick of the Santa Monica Mountains, the property overlooks the twinkling lights of the city.

The only access to the property, short of climbing the steep cliff, is a forty-degree turn off Cielo Drive onto a narrow road that twists its way to the gate of Sharon's house. At an acre's distance from the front porch, the closest neighbor couldn't really be considered next door.

A tick past midnight, it is quiet inside the house. Everyone has retired to separate rooms.

The living room occupies one-third of the floor plan. A fireplace and benched hearth spread across the back wall, windows across the front wall. A loft overhead canopies half the room; the other half exposed by a pitched roof encasing two dormer windows. At the base of the rafters, stereo speakers, softly playing music, looked over a piano, and next to the piano, a partner's desk.

Beneath the loft's edge, Woytek Frykowski naps on the sofa.

A hallway leads to two bedrooms. In the first, Gibbie, already in her nightgown, lays atop the covers of a Victorian bed with a book that doubles as a fan in the heat of the night.

In the other room, the added weight of the baby makes Sharon even hotter. She'd stripped down to her bra and panties, and then opened the doors leading to the pool. Jay sits on the edge of the bed, talking her down from an earlier squabble with Roman.

Outside, the temperature dropped a few precious degrees. Beyond the pool, a bounty of shrubbery sequesters the guesthouse. William Garretson lives there. Garretson and his acquaintance, Steven Parent, sit on the front porch sipping beers.

Beyond the opposite end of the property, a car with extinguished headlights rolls up the lightless road toward the gate. The passengers—Susan Atkins, Patricia Krenwinkel, Linda Kasabian, and Charles Watson—all concentrate on their objective and scout the area for anything that might hamper that mission.

There are two other houses on the road. When they pass the first one, the gate to Sharon's house comes into view. Watson parks at the fence. The neighbors don't hear the hum of the broken muffler even though their bedroom window is close enough that they should hear something.

Watson looks through the chain-link gate that serves not to hide, but to barricade the unwelcome; whereas a two-story garage and the swollen hillside near the fence obscure the residence and front lawn one hundred yards away. His neck cranes the height of the telephone pole before he jumps six feet to the first rung. At the top, he severs the phone lines and any chance of his victims calling for help.

The high view reveals the entire property. He waits. His eyes, already adapted to darkness, scan the grounds to see if the disconnected lines alerted anyone to their arrival.

Floodlights shine from the ancient oak trees. The branches sway and creak from the scurrying animals who found safety above the ground predators. Crickets tweak to one another over the rustling leaves. The wind alternatively flutters and surges through the canyon. Nature's ensemble shelters the single-story ranch, designed in the rustic fashion of French country homes. The red wood of the outer walls is illuminated here and there by coach lights. The front windows leak the interior's illumination onto the front lawn, shadowed by the criss-crossed panes.

Tex watches. Minutes pass. No one steps from the house to investigate the downed connection.

Back at the car, he shifts the transmission to neutral then pushes the decade-old Ford until it coasts to the bottom of the hill, where he obscurely plants it.

On their return to the gate, Watson coaches the women. They need to work as one to accomplish their goal.

At the same time, Garretson says good-bye to Steven at the door. Steven is left alone to cross the grounds to the driveway.

Garretson goes inside, turns up his stereo, and begins writing letters.

The engine to Steven's car turns over as Sadie, Katie, Yanna the Witch, and Tex climb the embankment to the right of the gate.

Steve turns the wheel of the drifting car to the left. He's searching for a control button to activate the gate.

The four killers drop down inside the fence when his headlights shine their way. "Get down and don't come out till I call for you," Watson orders the women.

Steven spots the remote mechanism. He slows near the pole.

Watson jumps from the bushes with a knife in one hand, a gun in the other, and forty feet of rope coiled over his shoulder like a Wild West cowboy. He levels the revolver until aimed at Steven. "Halt," Watson orders as he rounds the car to the driver's window. He lowers his face until he is eye to eye with the teenager.

Sensing danger, Steven throws the shifter into reverse. The knife slashes through the open window like a rabid dog's gnashing. Steven's hands jump off the wheel in defense, his foot slides from the brake. The creeping car veers toward the hillside. Shock motivates his silence when Watson's first swing of the blade splits the skin of his forearm, the second forceful enough to splice his metal watchband. "Stop hurting me." Steven pawed at the knife. "I swear I won't say a word to anyone, just leave me alone."

The razor edges cut through the tendons in his hands.

A bordering guardrail stops the car. Gunfire cracks. A bullet rips into Steven's arm. His heart thumps with explosive beats. "Oh God, why?" His bladder gives way, but he doesn't understand; only feels the warmth in the seat of his pants. Another bullet tears into his chest. His hand reaches toward the wound, but never completes the move; a final bullet slams into his cheek and through his skull; knocking him unconscious before his body sags onto the passenger seat.

Watson's hands bloodied, his shirt and face splattered, froze in the abrupt silence. He listens intently. Dogs bark throughout the canyon, but he can't tell from which direction they resonate. His alert eyes splinter the night, sweeping 360 degrees: toward the house, down the private road, then across the canyon, back to the house. The distant dogs settle; first one, then another until the dark hour is once again paralyzed.

At the end of the driveway apron, the property spreads out before them. Watson separates the group by sending Kasabian around to the back to look for open doors or windows. The other three go through a short gate and onto the front lawn. Ignoring the flagstone path, Watson trails the hedges bordering the front of the house. He checks the first two windows. Both locked. The third window, to the dining room, is raised a foot.

Kasabian returns as Watson cuts and removes the screen. He raises the window higher. One leg dangles over the sill. "Go keep watch out by the cars," he tells Kasabian. "If anyone comes, kill them."

"And listen for sounds," Krenwinkel adds.

"Shssh," Watson admonishes the remaining two. He hoists the other leg through the window. "Go wait for me by the front door."

Inside, he pauses to take in his surroundings. The floor plan is still fresh in his mind from his days visiting Terry Melcher. From the dining room he sees into the kitchen and the service area beyond it. The unlit rooms indicate they're vacant; he heads in the opposite direction, toward the lights.

The center of the living room holds a sitting area made up of two cream-colored brocade armchairs that serve as bookends to a beige velveteen couch facing the fireplace. In front of the couch sits a narrow coffee table, and in front of the table a zebra rug sleeps on the carpet.

The high back of the sofa conceals Woytek's slumber from the killer's viewpoint in the foyer.

The music plays in Watson's favor. The open-air loft can be accessed from a wooden ladder about a yard from where he stands. He goes four steps up to peer into the loft. Empty. On the way down, he realizes he's not alone. He studies Woytek's lethargic breaths. He lay on his back. His lips separated. Legs crossed at the ankles, one arm rests under his head, the other tucked into his waistband. His chest rises and falls to the beat of vulnerable dormancy.

Watson backtracks to the front door. When turned, the lock claps. He wheels his look to Woytek. Still asleep. He waves Atkins and Krenwinkel through, hushing their giggles.

Watson jabs at Woytek with the gun.

Groggy, he barely focuses on the three strangers. "What time is it?"

Watson kicks him in the head before quietly stating, "Shut up. One more word and I'll kill you."

Pain ringing in his ears, Woytek didn't hear the warning. "Who are you?"

"I'm the Devil, here to do the Devil's work," Watson hissed. Pointing to the hallway he tells Atkins, "There's two bedrooms down there, go see if anyone else is here."

Atkins first finds Gibbie, who looks over her book and smiles. After all, she is a guest and unalarmed at this odd voyeur. Atkins returns her smile, adding a wave for good measure before she moves on to Sharon's room.

Jay sits on the edge of the bed, his back to the door, his body blocking Sharon's eye-line. Atkins slips back into the shadowed hall unnoticed.

Watson used a bath towel to bind Woytek's hands at his back before shoving him belly up on the couch.

"There's three more back there," Atkins thumbs toward the hall.

Watson rolls his eyes. "Bring 'em out. Katie, go with her."

Armed with knives, they split at the bedrooms, Krenwinkel to Gibbie's, Atkins to Sharon's.

Atkins burst into the room, moving straight for her strongest adversary. She clenches the crown of Jay's hair and puts the knife to his throat before Sharon even has a chance to register the invasion. "Get up."

Jay slides off the bed while straining to see who is behind him. "Are you kidding? What's going on?"

"You'll find out soon enough."

Sharon shies against the headboard.

"You too," Atkins shouts.

Sharon edges out of the bed. "Can I just put on my robe?"

"You aren't gonna need it," Atkins said.

The three pause at the bedroom door. Ahead of them, Krenwinkel pulls Gibbie down the hall. She clutches the hem of her nightgown, and looks back at them, not quite frightened, but confused. Atkins pushes Jay forward then puts Sharon between them, the knife held to her stomach. "Move!" she orders.

In the living room, Watson sucker-punches Jay in the eye then whips him onto the chair closest to the hall.

Sharon hesitates at the archway. The easy familiarity of home has vanished. Her senses are heightened. The intruder's body odor is nauseating while her eyes are drawn to Watson's bloodied hands. She scans Woytek for injuries, searches his eyes for an answer to the festering storm.

With a scowl, Watson hurls her into the living room. Thrown off balance she falls to her knees.

"What the hell's the matter with you, she's pregnant. Be careful, dammit," Jay warns.

Watson bucks the gun against Jay's head. "Shut up or you're a dead man!"

Woytek shakes his head slightly, cautioning Jay.

Taking the rope he'd brought, Watson loops it twice around Jay's neck. The other end is thrown over the ceiling beam for leverage. Snatching Sharon, he loops the connecting rope around her neck.

Jay leaps from the chair. "I said to be more careful."

Watson rips the gun from his belt. "And I told you to shut up!" A discharged bullet mutes his words. Jay, shot in the chest, crumples to the floor. Watson dives on his back. He stabs him at least five times.

Sharon loses count. "Oh my God, what are you doing? Stop it! *Stop it!*"

Watson thrusts the bloody knife at her. "Shut up or you're next!"

A cry hangs in her throat. Silent tears tumble down her cheeks. In spite of the heat, her skin crawls with a chill.

Gibbie, trained to deal with gang members, knows to get to the crux of the situation before more violence erupts. "What do you want, why are you here?"

"We want all of your money," Watson demands.

"I have some in the bedroom," Gibbie offers.

Watson sends Atkins with her to retrieve the money.

Jay moans. His hands creep to his chest.

Sharon pleads with Watson, "Please help him, I think he's dying."

"You want me to help him?"

She nods.

He nudges Jay with his foot. "Hey there, buddy, you okay?" When Jay's eyes fluttered open, Watson makes a sweeping kick, cracking Jay's nose. "Is that better?" he asks Sharon.

Remembering the threat, she jerks with a wince, but doesn't make a sound.

Atkins and Gibbie return with her wallet. It contains seventy-two dollars. "That's it? Bunch of rich motherfuckers and that's all you've got!" Watson rants.

"How much do you want?" Woytek asks.

"We want thousands," Watson bellows.

"We just need more time," Sharon said. "We can get you as much money as you need. In the morning, you can take me to the bank. I swear I'll get you whatever you want just please don't hurt anyone else."

Watson paces. His anger intensifies with each step. Flipping off the lights, he turns with a vehement finality. "Forget it. It's over. You're all going to die!"

Woytek wrenches free of his restraint and lunges at Atkins. Striking her from behind, he tries to wrestle the weapon from her hands. She blindly flails behind her with the knife; one, two, three times she strikes Woytek in the legs. He screeches, falling to his hands when his legs give out. She seizes on his moment of weakness by driving the knife throughout his back, buttocks, and legs. She's a leech on his back while he crawls toward the front door. With a burst of energy, he throws her to the ground.

Watson watches the fight with amusement until he sees that Woytek is about to make it outside where his screams might reach help. Raising the gun, he fires a round into Woytek's left leg. A second blasts into his left rib, collapsing the lung. Doggedly, Woytek staggers to the threshold of the porch. The gun jams. Watson hurdles the couch; first tackling Woytek, then clubbing him over the head with so many forceful blows that the gun handle splinters around them. Woytek slumps to the floor.

Gibbie struggles against Krenwinkel, overpowering the murderess. Watson stalks across the room to take over. Slashing at her with the knife, he cuts the left side of her face, exposing her cheekbone. Another strike lacerates her jaw. The next plunges through her abdomen, stopping only when the blade hits her spinal column. She passes out.

Choking up blood, Jay steals Watson's attention. Watson kicks him in the head, then uses the knife across his back. The six-inch blade penetrates his lungs and heart. Jay wraps his arms about his head; his body shudders with gasping breaths.

Three feet away, on the couch, Atkins restrains Sharon's tangle for survival and clamps a hand over her mouth, forcing her to witness silently the murder of her friends.

Atkins catches sight of Woytek hobbling out the front door. "Tex, he's getting away!"

Woytek hangs on to a post on the front porch. "Somebody, please help me, oh God, help me!"

Watson races to silence him, his strikes as sure and swift as the needle of a sewing machine.

Like an apparition, Linda Kasabian appears from the shaded lawn. "Oh God, Tex! Why?" Ignored by him, she looks through the front door. "Sadie, make it stop, there're people coming!"

"It's too late!"

Woytek reaches out to Kasabian, his voice weakening. "Help me."

"I'm so sorry," she turns her back on him and runs.

Woytek tries to grab Watson's wielding knife, but both sides of the blade are honed, enabling it to slice through his hands like softened butter. Blinded by the blood flowing down his forehead, Woytek tumbles off the porch and through the hedges. He lands on the grass, with Watson on top of him, never missing a beat with the knife. Woytek's shouts blasts his ears. "No! Don't, oh God, please don't."

Inside, Gibbie regains consciousness. With her survival instincts intact, she runs down the hall, headed straight for Sharon's bedroom door. Before she reaches the poolside, Krenwinkel body-slams her into the shutters of the door. Gibbie yanks a clump of Krenwinkel's hair, pulling until her assailant loses her footing. With a second's upper hand, she pitches herself outside where her shrieks bounce back at her from the hillside. She's only taken a few steps before Krenwinkel tackles her.

Hearing the screams, Watson runs across the lawn to silence her. Like a cornered, weary child in a game of tag, Gibbie submits. "I give up, take me."

Watson's knife punches through Gibbie, along her right breast, down her sternum, over her belly, and into pelvic bone, continuing long after the life has passed from her. Satiated, he looks up and spots the distant lights of the guesthouse. Realizing his oversight, he sends Krenwinkel to make sure the house is empty.

Walking in the opposite direction, Watson crosses the front lawn until he finds Woytek straining like a kitten fresh from the womb, blindly seeking asylum. Watson stomps on his back, flattening him. He thrashes at him with the knife. So caught up in his assault that he doesn't notice when Woytek succumbs to death, continuing to propel the knife through the body over twenty more times. His arms tired,

Watson prolongs the attack by kicking him all around his torso and head. When he's finished, Woytek lay in a heap, unrecognizable as the man he'd been a half hour ago.

Moving toward the front door, Watson has forgotten that Sharon is alive; entering the house, her hysterics serve as a reminder. "Please don't kill me, please, I want to live, I want to have my baby, *please*, let me have my baby. *Please, I'm begging you*, take me with you. After I have my baby you can kill me any way you want. *Please, you've got to listen to me!*"

Atkins stoops until she's close enough that Sharon smells her sour breath. "Look bitch, I have no mercy for you. I don't care if you're gonna have a baby. Get ready, 'cause you're going to die and I don't feel a thing about it."

Watson steps inside. "Then why don't you kill her?"

Without hesitation, she swings the knife. Sharon flings her arms defensively, wildly, not knowing what she is hitting, and too terrified to feel the pain as the knife gashes her forearms.

Watson snaps the loose end of the rope looped around Sharon's neck. The velocity with which he pulls, suspends her, throttling her airway, leaving her defenseless when her fingers rip at the rope. Her widened eyes are her sole form of communication. They beseech, pray, even search for a godsend, but she is alone. Atkins stabs her near her left armpit.

Watson releases the rope and puts Sharon in a chokehold. She claws at his arms. "Dammit, Sadie, get her hands!"

Sharon writhes from his grip. As she twists away, his knife shot out a lick just below her rib cage. Her heart batters as she scrambles on her hands and knees, frantically crawling anywhere she can go. She reaches for the end table, grappling for leverage, but her blood-slicked fingers slip across the wood surface. His left hand catches her ankle and tugs her back; the other hand strikes out, tearing at her hamstring. A potted plant falls from the table. She launches it back at Watson. With the minuscule advantage, she scurries in the opposite direction, toward Jay, the rug burning her knees and elbows.

Only gaining inches before the entangled noose cinches her airway. Trapped, she curls her body around the baby; her hands wrap behind her head. Watson straddles her, his knife chops across her spine; his weight crushes the baby up toward her lungs. She struggles to hold onto consciousness.

Atkins crooks Sharon's right arm in her own and pulls to expose Sharon's chest. Watson buries the knife into her chest three successive times, puncturing the aorta and the sac around her heart. Blood pours from her chest, and trickles from her nose and mouth. She pulls her knees toward her chest, her breath labored, her final words as low as they are weak. "Mama, help me."

Watson and Atkins's heaving breaths ebb until the air is still.

"Are they all dead?" Atkins whispers.

Watson nods as he slumps to the couch, rolling Sharon's body out of the way.

Krenwinkel stands at the foyer. "There was no one in the other house."

"Are you sure?" Watson asks.

"Yeah. I looked through the windows, just a couple of dogs in one room."

Watson collects his weapons. "Let's get out of here."

They make it to the driveway. "Shit," Watson stops them, "we forgot to write stuff on the walls. Sadie, go back while we get the car."

In the living room, Atkins takes the towel that had bound Woytek's hands before kneeling by Sharon's body. The blood continues to seep from her bosom. Atkins's fingertip swirls over Sharon's skin like a finger painter. Then she puts her bloodied finger to her lips. Her tongue reaches to taste it. Sitting as if in a trance, her other hand goes to Sharon's abdomen. The baby is probably alive. Should she cut it out? Wow, what a trip, she thinks, to taste death, and yet give life.

Watson is back at the front door. "Sadie, hurry up!"

The idea abandoned, she puts the towel to Sharon's chest. The thirsty tip soaks until it is maroon. With the towel, she writes PIG across the outside of the front door.

Atkins caught up with them at the driveway. They recover their unsoiled clothes from the bushes. This time Watson uses the control button to open the gate. As it opens, the three killers escape, disappearing into the night, where they remain undetected for four months.

I CLOSED BUGLIOSI'S book. Contrary to my expectations, it didn't have all the answers. It was missing emotion. How did victims *feel* in their last hour? Any thesaurus bulges with insightful compilations; however, each is an inadequate depiction of what that emotion must be. Perhaps we don't need to invent an appropriate word because a victim's instinct for survival overrides emotion. If not, I doubt the human language will ever succeed in finding a suitable phrase; after all, the emotion is caused by the inhuman.

And what do the inhuman feel when they kill? Hatred, revenge, outrage, fury, injustice, dissatisfaction, jealousy, greed? Was the mere action of stabbing a catalyst that propelled them to annihilate?

In place of a knife, I took a pen and stabbed at the ottoman. *One, two, three, four*—my eyebrows naturally furrowed by the action. *Five, six, seven*, my jaw clenched. *Eight, nine, ten*, the arc of my swings heightened. *Eleven, twelve, thirteen*, an unidentified anger brewed; the very act enraging. *Fourteen, fifteen, sixteen*, I paused to mark the effort it took to kill Sharon. *Seventeen, eighteen, nineteen*, adrenaline rushing, I stood in order to stab harder. Continuing on fifty-one more times as in Woytek's death, I hit at a frenetic pace. Twenty-eight times for Gibbie's murder, and so on until my arm swung one hundred times. One hundred one, one hundred and two, I dropped the pen.

A rush of air escaped as I slunk into the chair, kneading my wrist and elbow. I concentrated on what I felt.

Release.

During the past eight hours of reading depictions of murder, anger had replaced fear. I was damn good and angry with Manson, Watson, Atkins, Krenwinkel, Kasabian, and Van Houten. It was venom that needed to be released, and the experiment served the purpose.

Limp against the cushions, I felt much like the surrender of a good long cry, exhausted and numb. The exact words Atkins had used: "Tired, but at peace . . . exhausted."

It took a tremendous effort to reenact the stabbing, with an innate crescendo of anger building with each swing. A meager depiction of the fury. Still, it gave me the understanding of how personal it is to stab someone. Different from other killing breeds, the stabber has to look into their victim's dying eyes. They feel the victim's quivering skin while holding them at bay. The stabber's hands are one with the weapon; striking until they feel the last breath, their hands and arms soaked with blood by the time they're finished.

This is the type of murderess Susan Atkins is.

In all that I'd read about Atkins, a quote from the book *5 to Die* is the most insightful paragraph ever written about this woman. The authors, Jerry LeBlanc and Ivor Davis, interviewed Atkins before the trial when little was known about her.

"At first, she said Gary Hinman had been cut up in a fight and she went to his house to nurse him. . . . When it was inferred, falsely, that Bobby Beausoleil had already fingered her, Sadie loosened her tongue and let fly. She hasn't stopped talking yet, but always, she tells just as much as has to be told, and, it will be seen, in whatever version suits her passing whim. . . . A facial chameleon whose features can shift almost imperceptibly from impatience to engrossment, from unconcern to rapt attention, from irritation to mirth, and from coldness to glowing innocence. If Atkins goes free, it just might be by the recitation of so many divergent, unlikely stories, always within a fabric of truth, that a jury of her peers, if any exist, will have to conclude that either she or they have lost the power to distinguish between fantasy and reality."

Between August 1969 and January 1971, Atkins's version of whether she stabbed Sharon or not, alternated with a seesaw's pace. Fact or fiction, the most disturbing aspect of her *many* confessions is that her mind could even formulate the hideous details she provided.

Before and after her arrest for the murders, Atkins boasted to anyone within earshot that she was the one who'd stabbed Sharon. In his book, Vince Bugliosi quoted Atkins's confession to a cellmate before her indictment for the murders. " 'Sharon was the last to die.' On saying this, Susan laughed. 'I proceeded to stab her. It felt so good the first time I stabbed her, and when she screamed at me, it did something to me, sent a rush through me. . . . It was like going into nothing, going into air. It's like a sexual release, especially when you see the blood spurting out. It's better than a climax.' "

By the time she testified for the grand jury, her account changed. "In order to make a diversion so that Tex couldn't see that I couldn't kill her [Sharon], I grabbed her hand, held her arms and then I saw Tex stab her in the heart."

During the penalty phase of her trial she reverted back to an earlier version of her statement. "I was still holding Sharon Tate and Tex came back in and said kill her and I killed her. . . . I stabbed her. . . . She fell and I stabbed her again. . . . I don't know how many times I stabbed her. She sounded like an IBM machine. She kept begging and pleading and pleading and begging, and I got sick of listening to her, so I stabbed her and I just kept stabbing until she stopped screaming."

In 1972, when California removed the killers from death row, Atkins did another about-face to exonerate herself of Sharon's murder.

In 1976 Atkins metamorphosed into a born-again Christian and cleansed her soul. "The whitest, most brilliant light I had ever seen poured over me," she wrote. "Vaguely, there was the form of a man. I knew it was Jesus. . . . He spoke to me. . . . 'Susan, I'm really coming into your heart to stay. Right now you are being born again and you will live with me in heaven through all eternity. . . . You are now a child of God. You are washed clean and your sins have all been forgiven. . . .' There was no more guilt! It was gone. . . . The bitterness too. Instantly gone."

Ten years and several parole hearings later, Atkins gave her first television interview since the trial. The *60 Minutes* program opened with an image of Atkins in prayer before a portrait of Jesus. A voice-over from the host touted that she was about to grant viewers "an extraordinary new version of the killings."

Atkins tearfully explained, "I've spoken the truth as I've talked to you and the truth is I never killed anybody. I lied at the grand jury. I lied at the trial and said that I killed people that I had in fact not killed. I live with the fact that people think I killed a pregnant woman."

"Would you clear up once and for all what actually happened that night?" the reporter asked.

"I was instructed to go to the Tate home. I went to the Tate home, and I saw people brutally murdered, and I was instructed to kill two people, and I could not do it. Everything I was told to do that night I did with the exception of killing somebody . . . it was terrifying, horrible . . . could we not talk about this? It's really hard."

Curious if she had changed her story since *60 Minutes* aired, I turned to the record of her 1987 parole hearing, the one following that TV interview. Not surprisingly, she'd added another twist to her story. "I want once and for all for it to be understood that what I have been accused of doing and what I actually did are two differ-ent things. I'm asking that you believe me when I tell you I did not by my own hand kill any human being. I said and spoke several differ-ent stories between the grand jury and the trial that were in fact not true. What I said at the grand jury *was the truth*. I ask for your indul-gence that if you were to read the coroner's report and police reports, you'll see that all of the stab wounds to all of the victims were made by the same weapon. And it was not the weapon that was found at the crime scene, which was my knife that I lost while struggling with Mr. Frykowski, and the forensic tests showed that there was no blood on that knife."

Willing to indulge Atkins's theory, I read each transcript she'd men-tioned. "He (Frykowski) got behind me," Atkins testified for the grand jury, "and I had the knife in my right hand. I was just swinging. . . . I remember hitting something repeatedly behind me. I didn't see what it was that I was stabbing . . . it could have been a chair."

I read the crime scene report and scanned the photos with a mag-nifying glass. There was not a single rip in any of the furnishings, not a single gouge in the wood surfaces, not even a tear in the carpeting.

In truth, the coroner's report attested to Atkins's stabbing Sharon.

In preparation for an eventual trial, Dr. Noguchi wanted to have the exact dimensions of the knives used in case they were recovered. To do this, he poured barium-sulfate paste into each wound and then took X-rays. Sharon's wounds ranged in size from some that were 1 inch in width and 3½ inches in depth to others that were 1¾ inches in width and 5½ inches in depth. He concluded from those tests that two knives were used, one smaller than the other.

Testimony from the four killers supports that Patricia Krenwinkel was out of the main house when Sharon was murdered; that leaves Watson and Atkins as the two knife wielders.

Whether Atkins actually stabbed Sharon has no bearing on her culpability. The pertinent issue is this: If she's lying about her participation, then she can't possibly be rehabilitated or remorseful.

In 1985 Mom attended Atkins's parole hearing, giving Atkins the chance to extend personally her condolences. Instead, Atkins emphasized a need for absolution.

Over the next four years, Mom made Atkins's lack of remorse a point of contention, goading her into saying, "Mrs. Tate, I am sorry." Nevertheless, at her 1989 parole hearing a board member asked, "Miss Atkins, who have you made amends to?"

"To God, to myself, to my family. I don't think that the family members of the victims are willing to make amends, though I would like to do that. Sometimes I cry because I see this woman (Mom) sitting over here in tears and I know her anguish is because of actions that I did. I feel their pain every single day." She went on for twenty minutes about her emotions, but noticeably neglected to say those five simple words.

The last bit of information that I wanted to explore was an Internet page that Atkins established with the help of friends on the outside. The website appeared to be dedicated to "spreading the word of the Holy Spirit." Atkins wrote many prayers on this site, for herself and for other inmates—especially those on death row. It includes thoughts about God, Christ, and religion, and sentiments shared on her salvation through repentance and rehabilitation. What's unmistakably missing from her wisdom are the victims. There is not one prayer for my family, not one word of remorse directed to the eight families of her victims.

The final page of her site held one last prayer—for her parole hearing: "I've prayed much about this after I heard from the Holy Spirit on it. I write it only at his leading and direction. I will once more appear before the California Board of Prison Terms for the purpose of determining whether or not I am suitable for parole. This letter is from my heart, to ask you to pray with me that only God's will be done when I appear again. That He will fill my mouth with His wisdom that day . . . hope-filled loving sister in Christ, Susan Atkins."

18

CHANGES

We were each a different person every day of our life. In fact,
we were three, four, five different people. We changed clothing,
changed expressions. Just changes. You know, changes are changes.
They're not progress, they're not regression. They're just change.

—SUSAN ATKINS, 1971

Patti

Mixed between palpitations, my heart shivered in protest for the entire three-hour drive to the prison. I was on my way to meet the bogeyman and scared to death. I pulled off the lightly traveled highway
a half-dozen times when my conviction faltered.

Parked on the shoulder, I felt as if I were in one of those old cartoons
with an angel on one shoulder and the devil on the other, bantering
back and forth in my mind. "You can't break a deathbed promise to
your mother," the angel coaxed.

"Sure you can, she won't care anymore," the devil assured.

"Sharon did so much for you, you must do this for her," the angel counseled. "Remember the picture of her, crumpled against the
couch? She needs you."

"Yeah, remember that picture—because it could be you if you go through with this," the devil warned.

Before the men in white coats could show up with a straitjacket, I pulled back onto the road, just trying to take it mile by mile until I saw a sign that indicated the California Institute for Women at Frontera was the next exit. My coffee cup was empty, my throat was parched, and my eyes stung from salty tears. I'd crushed one empty pack of cigarettes and opened another before I pulled into the prison parking lot and took the first available spot.

I draped my shoulders across the steering wheel, feeling more like I'd run the distance than driven it. When my breath slowed to a steadier pant, I chanced a look in the rearview mirror to find a face I hardly recognized. My makeup had been wiped away hours ago from the steady flow of tears, enhancing my sleep-deprived eyes. Jittery hands had matted my hair and spilled coffee stained my blouse. I shook my head wearily. I'd fooled myself into thinking I could confront Susan Atkins. I'd studied like a good student for the final exam, but in this case, knowledge didn't bring power, just more fear, anxiety, and discouragement.

I was in no condition to attend this hearing. The parole board would see right through my weakness, and any persuasion Mom may have had over them in the past would be instantly undone by my instability. Conceding defeat, my head returned to the steering wheel for a nap before I'd turn around and make the trip home.

Sleep came instantly, but lasted only seconds before there was a tap on my window. Disoriented, my blurry eyes focused. Oh, no, I'd completely forgotten that Kelly was attending the hearing as my support person. She was also the executive director of the Doris Tate Crime Victims Bureau and had taken me under her wing to guide me through the bureaucratic complexity of victims' legislation.

The look on her face said that my escape wasn't going to be easy. Still seated in safe surroundings, I opened the door. "Look, I can't do this, Kelly. I'm sorry you drove all the way out here."

She handed me a fresh cup of coffee. "Yes, you can."

"No, you don't understand."

"I don't understand?" She smiled. "I just went through this for my brother. Now let's go."

My only move was to take a sip of coffee.

Her look hardened. Though ten years my junior, she had Mom's stern force. "Patti, this is the point in your life that you stop being the victim and start being the fighter. Now, you're going to get out of that car, put up your dukes, and go get that witch."

My first parole hearing didn't have the celebrity of Mom's. Purposely low-key, I filled out the prison attendance forms using my married name. There were no petitions to deliver and no press conference to be held beforehand. Only the press's speculation about how I was connected to the Tate family. And so we easily walked through the thick of the press and my last steps of sheltered anonymity. Steve Kay waited in the lobby. "I was about to give up on you," he said. "Come on, they're waiting for us and not too happy about it."

A prison guard led the way to the dark paneled boardroom. I kept my focus telescoped to the ground and took the seat next to Steve at the head of a T-shaped table. My rear end barely hit the chair before the board member sitting closest to me pressed the RECORD button on a tape machine. "We're on record. This is a subsequent parole consideration for Susan Atkins, I'm Chairman Manny Guaderrama. Miss Atkins, the way this hearing will be conducted today is that I will be discussing your life offense, your prior criminal and social history. And then Miss Bentley on my right will discuss post-commission factors, and your adjustment in the institution. Then Mr. Neilson on my far right will discuss parole plans.

"You are not required to admit your guilt," Guaderrama continued with hardly a breath, "but the panel does find true the findings of the court. You came into this room convicted of murder and you'll leave this room convicted of murder. Our charge here today is to see if you are suitable for parole.

"Okay, now what I'm going to do is read a statement of facts from the hearing held on July sixth, 1979, your initial hearing. Referring to case number A267861, the victim, Gary Hinman, was apparently cut and stabbed repeatedly. Case A253156, referring to counts one through five, these concern the murders which occurred on August ninth, 1969 at the Polanski residence. . . ."

I felt Atkins gaze settle upon me; it caused my heart to hammer so hard that I thought it would bounce out across the table and right into her hands. Perspiration beaded across my hairline. My underarms burned with the surge of nervous sweat. My goose bumps–plagued arms banded across my stomach as it lurched in protest. Too many cigarettes, too much coffee, and not enough sleep. My head throbbed to the rhythm of my pulse. The room swirled. Oh God, I was going to vomit, and if I were going to toss my cookies on national television, I may as well be looking at her. My eyes settled on the bogeyman—or woman, as it turned out, and an old one at that.

Atkins's dress was short-sleeved and innocent-pink. Except for the sides, which were clipped back, her graying hair fell below her breast. Oversize smoked lenses that she awkwardly pulled off and put on masked her black eyes. Her red nails were long as daggers, and her ruby lipstick had bled onto her upper teeth like the vampire she once claimed to be. Our eyes met briefly before she looked away, and I realized she was powerless. She had come to plead for her life much like her victims had pleaded for theirs, and I had come to keep her quashed in a lifeless prison. My stomach and heart settled at once.

"These murders have been gone over a number of times," Mr. Guaderrama was saying, "is there anything that you would like to say concerning your involvement, Miss Atkins?"

"Yes. There is no way to justify or excuse my participation. I would also like to state that it is impossible to explain to intelligent people insane actions that occurred twenty-three years ago from a cognizant, aware position. And that's what I was living in, insanity."

Atkins's words came slowly, as if addressing children, yet with a rehearsed eloquence. "I left home at eighteen. I came from a very, very dysfunctional family—again, it's not an excuse, just a statement of fact. I had two alcoholic parents. I suffered sexual molestation as a young child. When I left home at eighteen, I was extremely angry, extremely vulnerable, and directionless. Couple that with my involvement in drug abuse and my need to be loved and accepted, I became very susceptible to the times, and I ask that the panel members take this into consideration—not as an excuse—but just to take it into consideration.

"When I met Charles Manson, he was extremely charismatic and I was in love with him. My intent when I met Charles Manson was not to spend the rest of my life in prison. There was a combination of circumstances that brought me to the homes of the people that died. Again, there's absolutely no excuse. I take absolute and complete responsibility for my actions, but I did not by my own hand kill any human being. I was there, I participated, I did nothing to stop it—and that's inexcusable, but I did not raise a knife to anyone.

"There were several things that I've said that were not true. During the trial, I was asked by Charles Manson to take the heat off him and my co-defendants and so I stated that I had stabbed Sharon Tate, when in fact, I did not. It almost seems pointless to say this—"

Guaderrama interrupted. "Let me ask you, what *were* you doing while all this killing was going on?"

Atkins took a deep breath. "When I was told to get in the car and go with Tex Watson, I did not know exactly what was going to happen. I was told by Mr. Manson to do what Tex said to do. So I got in the car and did *exactly as Tex said*. When he told me to climb a fence, I climbed a fence. When he told me to hide in the bushes, I hid in the bushes. When Steven Parent was killed, I was watching. When I was told to go into the house, I went into the house. I was told to go into the back rooms and see how many people were there and I did that. I was told to tie up Woytek Frykowski and I tied him up with a towel. And I was told to kill him, but I was incapable of doing it. I had raised the knife to do it and I could not do it. At that point, Woytek Frykowski jumped on me and we got into a struggle. I screamed for help and Tex Watson came over and took him off me. Tex then screamed at me to watch the other woman, and that happened to be Sharon Tate. And I sat in front of her until Tex came back."

"Were you holding her hostage?" Guaderrama asked.

"No. I was just sitting in front of her."

"Did she try to get away?"

"No, she did not. I believe she was tied up."

"Did she say anything?" Guaderrama asked.

"Yes."

"What did she say?"

Atkins closed her eyes before answering and I wondered what went through her mind; did the murders play out across her lids or was she conjuring a way to keep out the conversation between her and Sharon? "Oh God," she said tearfully, "she asked me to let her baby live."

"What did you say to her?"

"I told her I didn't have any mercy on her—" Her tears conveniently cut off the rest.

"Was she physically hurt at this point?"

"No. Just tied up."

"And the others? Where were they?" Guaderrama asked.

"Abigail Folger had run out of the house. Jay Sebring was lying on the floor. He had been shot."

"Hmmm," Guaderrama fished out a piece of paper. "I was curious, in your probationary report, the evaluator comments: 'The defendant stated with some bitterness that she was not even at the LaBianca killings, nevertheless she was convicted. The defendant claims, "I killed Sharon Tate. I don't know why I killed her. I was on acid; whenever someone tells me not to do something, I do it. She said don't, so I did." ' So what I was curious about is why don't you take responsibility for killing the LaBiancas if you take responsibility for killing Sharon Tate? I mean, you've very emphatically denied that you had any involvement in killing the LaBiancas—"

"Because I never entered the LaBiancas' house. I didn't kill them," Atkins said.

"Yes, but you were an accomplice. You didn't stop the murders and in this case, today, you just told us that you didn't kill Sharon Tate either. I don't understand why you take credit for her murder, but not for the LaBiancas."

Atkins squirmed in her chair. "See, when I made that statement twenty-two years ago—I don't know if you can understand this—"

"Try me," Guaderrama said.

"Okay. When I was first arrested for the murders, I had a lot of guilt for what I had done, but I didn't want to admit it. I could only hear Charlie in my head saying, 'When you're in jail, you have to be big and mean so that everyone will leave you alone.' I was very frightened and I had no support mechanism around me. All I knew was that I somehow had to convince everybody that I was mean so I wouldn't die in jail. That's why I said what I said and that's why I didn't take responsibility."

"But you still don't—never mind," Guaderrama waved the thought away. "I'm going to move on to your counselor's report that was prepared for this hearing. Your counselor, Mr. Salcedo, indicates that your behavior has essentially remained the same since your last hearing in 1989 . . . He does indicate, however, that based on the seriousness of your commitment offense and the nature of your crimes that you continue to be a high risk and a threat to society if released at this time."

I leaned toward Steve and whispered, "Does that mean we can go home now?" He smiled, but I was serious. What in the world were we all doing here if her counselor deemed her a threat?

We'd been in the hearing for three hours. I escaped for a few minutes to reflect on Sharon's life rather than her death.

I closed my eyes and I was a child again, waiting with Mom for Sharon to step off the plane from New York. She'd been in Italy for three months filming *12+1 Chairs*.

In my collective memory, Sharon's appearance had never changed during my childhood: blondish-brown hair, thin, and beautiful. When she appeared on the gangplank, I ran to hug her, but something had changed this time—she was as fat as a horse! I gaped at her belly. "What happened to you?"

Her eyes shined. "It's just the baby, silly. Here, give me your hand, you can feel her moving around." My hand rested flat on her stomach. Nothing. Then, with a quiver, her belly came to life. I jerked my hand back. "Did you feel it?" she asked. I nodded while hesitantly reaching back to feel it again. Her arms reached out to pull me close for a hug. "Don't worry, you're not going to hurt her," she laughed. "When we get home, you can lay your head there and hear her heartbeat."

By the time we got to her house, Sharon was exhausted. "Mom, I've got so much for the baby that I want to show you, but I've got to lay down for a bit."

"Go ahead," Mom told her, "I'll start unpacking for you."

Sharon started down the hall. "Patti, do you want to come with me?"

I looked to Mom for approval, and then followed closely behind, snuggling right up to her on the bed. I lay at her side with my hands on her belly waiting for the baby to move. "Here," she said, putting a pillow right next to her stomach, "lay on this with your ear right next to the baby. Close your eyes and be real quiet."

It took a moment, then, as though far away, there was an echo after Sharon's heartbeat and then another. Captivated by the sound of this new life and awed that my sister had created it, I could have stayed like that forever, frozen in time, never in need of a better moment. I looked up to find a single tear sliding down her cheek. "Why are you crying?"

She smiled. "Don't worry, it's a happy tear. I was thinking about the day you were born. From that moment on, I couldn't wait to have my own baby."

Although it was a scorching day, I moved in closer to the comfort of her bosom and the rhythm of her breaths that had so often lulled me to sleep during my childhood.

When I opened my eyes in the boardroom, I could still feel Sharon's warmth and missed her more in that instant than I had in all the combined memories over the years.

"Very last question," board member Neilson said. "Miss Atkins, who are the victims of the crimes that you participated in?"

Atkins sighed heavily. "Every person that's suffered, particularly and beginning with the family members of the victims. They are the first victims and the most innocent victims."

Neilson frowned. "What about the ones that you killed? Sharon Tate's baby for instance? Aren't they the first victims?"

Atkins eyes moved quickly across the panel members. "Yes, of course. . . ." Her voice trailed off, and for first time in three hours, she was speechless.

"Miss Bentley, do you have any more questions?" Guaderrama asked.

"Yes, just a few. Were you on any trips to scout the crime scenes?"

"No. I was not."

"So, all of the sudden, you were told one night you were going to do these murders?"

"Well, I didn't know that we were going out to murder anyone—"

"What were you told you were going to do?"

"I wasn't. I was told to get a set of dark clothes and a knife and to go with Tex Watson and do whatever Tex Watson said to do. And that's what I did. And I had a sick feeling."

"And this happened to you on both nights, this sick feeling?"

"On the night of the LaBiancas I did not want to go. And I don't offer it as a justification, but it is part of the facts that I was very, very frightened that if I did not do what I was told to do, I would be hurt."

"That you would be hurt? Was that more important—"

"Well my son's life was threatened—"

"That's all I have, Mr. Guaderrama," Bentley told him.

"At this time, we'll ask Mr. Kay to give a summary on parole suitability," Guaderrama said.

Steve sat forward in his chair. "It's frustrating sitting here listening to Susan Atkins. She said today that she holds the truth dearly. That she could never face remorse until she faced the truth. She has lied to you so many times today about her participation in the crimes that I lost count.

"During her life, Susan Atkins has been involved in four significant events that the board needs to concentrate on, because her involvement in these events has indelibly categorized her as one of the most dangerous female criminals in American history.

"The first was her arrest in the state of Oregon in 1966 when she was eighteen years old—before meeting Manson. How she described this incident to you is completely flower-coated as she embellished herself as a hostage of the two men she was arrested with. She was an *accomplice* to these two wanted men for grand theft and the Dyer Act.

"When the Oregon state trooper who arrested her testified during the penalty phase of the trial, he said that she was carrying a loaded gun and upon her arrest she told the trooper, 'I should have killed you.' Now, in 1966, I think you would have been hard pressed to find teenagers carrying a loaded gun, let alone one that would threaten an officer of the law.

"The next serious event she was involved in was the murder of Gary Hinman. Now, Miss Atkins and her crime partners tortured him to death over two days. At the penalty phase of the Tate-LaBianca trial, Atkins admitted that she stabbed Mr. Hinman. And at the Hinman trial, she admitted that while he was dying, she held a pillow over his face to suffocate him.

"On August eighth, she left with her crime partners for Sharon Tate's house. Now, she told you today, 'I didn't know what was happening, I had a sickening feeling.' But one of the ranch hands, Juan Flynn, testified at her trial that he saw the group leaving and asked where they were going, and Atkins leaned out of the car window and said, 'We're gonna kill some motherfucking pigs.' She knew what she was going to do—she left the ranch armed with a knife—make no mistake about it.

"After the murders, she spent the next week at Spahn's Ranch sharpening knives, and telling people how sexual it felt to stab someone—"

Atkins's attorney interrupted. "I'm going to object to this entire rendition because it's coming from the third trial of Leslie Van Houten—"

Atkins rose inches from her chair. "Yep, that's right, this was not at my trial."

"Yes it was," Steve said.

Simultaneously, Atkins's attorney said, "It was not."

"Hold on," Guaderrama held up a hand. "It's not in the file, Mr. Kay."

"It is in her file. The facts don't change—Miss Atkins's story might change—but the facts remain. These victims were literally mutilated and Miss Atkins participated."

"Let's move on, Mr. Kay," Guaderrama said.

"The fourth event was of course the LaBianca murders—and oh, is she ready to go out again. Now, she didn't go into the LaBiancas', but she was convicted as a coconspirator of those murders.

"In conclusion, Miss Atkins has been convicted of eight murders. Consider how many lives that is." Steve scanned the room, counting heads. "We could eliminate—outside of Atkins—everyone in this room, all of us gone—poof. With that in mind, I ask you to find her unsuitable for parole and deny her for the maximum five years."

Guaderrama turned to Atkins. "Is there anything you'd like to say that hasn't been covered by your attorney concerning your suitability for parole?"

"Yes," Atkins replied with welling tears. "Most people that don't know me put twenty-three years of history aside and take three nights—three horrible, horrible, horrible nights—and superimpose that on me, and then view me only through those nights. I'm asking you to remove that superimposed image and look at just me."

Atkins glanced at me and jump-started my heart with a wink and a smile that lasted only as long as the blink. I looked to see if anyone else noticed. They appeared oblivious to the incident. When I looked back at Atkins, she was looking toward the board members.

"People say I have no remorse. I have remorse. Twenty-three years of incarceration, twenty-three years of therapy, twenty-three years of self-help groups. I couldn't go through that unless I felt remorse. I was driven by remorse to try to find answers as to why I did what I did and to ensure that I never repeat those kinds of offenses again. I couldn't do those things and still be the crazed killer that Mr. Kay presents to you today."

Atkins dabbed at her eyes. "I could not put on twenty-three years of acting. Mr. Kay said three years ago, in front of the television cameras, 'We know she's going to cry, we just don't know when. It's all an act.' That's not true. I cry because that's how I feel and I *feel* every day. That's all I want to do today is feel and love, and care about people. My life in prison is dedicated to that. I do it every day when I get up. I ask God to give me the grace—"

"Could we keep your comments to your suitability for parole?" Guaderrama interrupted.

"That's what makes me suitable, my twenty-three-year history here. The crimes in and of themselves—there is no way they would ever make me suitable for parole. But this is my new social history, it's in this institution and I think it speaks very well for itself. I have a viable marriage. The first healthy relationship I have ever had, and I cherish it, and I take care of it, and I nourish it. It's the most precious gift I've ever been given.

"Whatever Mr. Kay's rendition of the crimes were, I was the only one in this room that was there. I am the only one that has to live with it every single day—aside from Miss Tate, who is here—"

"Sum up your remarks," Guaderrama censored.

Atkins's tears flowed heavier. "I ask you to consider giving me a second chance. I am the least deserving of this, but I have the courage and conviction of my heart that I sit before you a successful, rehabilitated person."

"Thank you, we'll consider your remarks." Then Guaderrama turned toward me. "At this time we're going to turn the floor over to Miss Tate. Would you like to say something?"

"Yes, I would." I looked to my hands and saw that I had adopted Mom's habit of shredding tissues. I swept the pieces up into my palm and prayed for her guidance. "You know, during all this, it takes me back into my childhood memories. I was just eleven when this happened and as far as I was concerned, we were on top of the world until August ninth.

"The last time I saw my sister at her house I was so excited about being an aunt and she was resting because she was so close to her due date and I used to sit with my hands on her belly by the hour trying to feel that baby kick. And I was so proud of my sister. I mean, we had it all, what more could the good Lord give us? To me it was just the best. I had a sister that I looked up to. Our family was good and we were glad to have Sharon home. She hadn't even been home that long, a couple of weeks maybe, and I spent every second I could with her.

"We had good times before they killed her.

"When I left my sister's house one evening, the last time I saw her, life was good. And then I woke up another day and life had changed so very, very dramatically, to never be the same again.

"I lay by my mother for months, wondering if my mom was ever going to pull through this. I'll tell you, for years, *years* life was so hard, to get by day to day. Still, life is hard. This was such an evil, evil thing, and for a child to try to have to soak it in, it didn't make sense to me then and it still doesn't make any sense to me. I was so frightened and I have lived with fright *every day* of my life since. I look over my shoulder every day, wondering if I'm safe.

"Life is very uncertain and life is very uncertain outside of this prison. Life is not easy on the outside, not for me, not for anybody. Miss Atkins refused to live in our society and shunned society. Well, that's just not the way it works. You know, we all have to obey the laws, the laws of God and man. It worries me to think of her outside this prison. I don't see how people will accept her on the outside. I'm not ready for murderers like this to live next door to me. This whole situation still frightens me very much.

"What I'm saying, in essence, is that I watched a big part of my mom die when they killed Sharon. She suffered so horribly for the rest of her life. What I'm asking is that there is a consequence to be paid here for all the destroyed lives, and that's with Miss Atkins's life in incarceration. I believe that even though she may be rehabilitated—and it sounds to me like she may be doing very well—her work here in the prison is very needed. She can do a lot of work right here and I believe that right here is where she should stay for the rest of her life."

A half hour later, the board announced their three-year parole denial. I looked at Atkins to find her staring at me. Her vengeful glare only gave me more strength. I held her stare and said, "God's will be done."

I felt drunk with energy now that the ordeal was behind me. "You made it," Kelly said. "I'm so proud of you!"

"I'm proud of me, too," I admitted. "You're going to think I'm crazy, but Mom was there. I could feel her all around me, giving me strength."

"Well, I hate to bring you down, but the press is going to want to talk to you now."

"It's okay. I'm ready for them."

Outside, Doug Bruckner from *Hard Copy* approached me first. "Miss Tate, how did it feel attending your first parole hearing?"

"It was hard going in there and looking at the last face that my sister saw, but I felt okay. It brought up a lot of old memories, and that was hard to deal with."

"How did you feel when you found out that she would have to remain in prison?" Bruckner asked.

"Well, actually, I kind of hoped for five years. The five-year denial is something that my mom fought hard to have passed through legislation, and I would have liked to honor her memory with it, but three years is fine."

"Did this experience change your mind about attending the hearings for the other killers?" he asked.

"No. It gave me courage. I'll be at Watson's later this year, and before his hearing I'm going to begin a new petition drive, same as my mom's ten years ago."

Another reporter approached. "Patti, what was your reaction when

Atkins said that she felt your pain and understood the anguish that you go through and that it proves that she's rehabilitated?"

"Actually, my feeling was that it didn't really matter. She committed a terrible, terrible crime, and she has to pay for it whether she is rehabilitated or not. Is she remorseful and feels my pain? I don't know and I'm not going to judge, so that part really went right over me."

"Did you look at her much of the time?" the same woman asked.

"Yes I did, and I listened to everything she said. But what she says is so abstract. I'll never understand why they did what they did—even Atkins says that. There's no rhyme or reason. It's just the way it is."

"Do you believe there's a chance that any of the Manson Family members will ever be released?" A voice in the crowd asked.

"You have to remember that the parole board once gave Sirhan Sirhan a parole date, and that's about the time that my mom stepped up to the plate to stop this kind of nonsense. So, yes, there's always a chance."

Steven Kay edged up beside me. "Steve," the reporter asked, "do you believe that Susan Atkins is rehabilitated?"

While he answered, I took the opportunity to slip back from the bright camera lights, but I heard his response as I made my escape. "Absolutely not. This is one of her worst performances in the last five hearings. She blatantly lied about her participation in these murders and was not willing to take responsibility. The evidence against her was clear at the trial and here she is today blaming everyone else. So I think that shows that she's not rehabilitated in the least."

19

MANSON AND A ROSE

I naïvely thought that there was a certain dark humor in Charles
Manson singing love song lyrics, but now I find the word "humor"
doesn't fit into the equation at all. Especially when I think about
the families of his victims, and how it makes them feel.

—SLASH, GUITARIST FOR GUNS N' ROSES, 1993

Patti

"Mama, come check this out," Brie called from the family room, where music blared from the MTV channel that my kids watched like Valium-overdosed patients, happy and drooling.

On the screen, a man sang to thousands of fans during a concert. Staring wildly from the singer's shirt was a photo of Charles Manson. I flipped off the television's power.

"What are you doing?" my eleven-year-old daughter asked. "That's Axl Rose! He's so cute." She grabbed the remote and turned the television back on.

"Hey," I protested, but then I decided to watch what Brie found so attractive about the lead singer of Guns N' Roses. When the song ended, I turned the volume down. "Hon, he's wearing a Manson T-shirt. What message do you think he's sending out to his fans?"

"I guess that he likes him."

"Do you think that's okay?"

"Not really."

"What makes Axl Rose so cool?"

Brie shrugged. "I don't know."

"Do you like his music? The lyrics? What?"

Another shrug. "Jeez, give me a break, I don't know, he's just cute."

So much for insight.

I tried to keep an open mind about the rock groups that tantalized my kids. After all, my generation had Alice Cooper, Black Sabbath, and Steppenwolf, all bands that were thought of by some to be evil and a corrupting influence on our youth. Nevertheless, by sporting Manson on his chest, Axl Rose opened himself to my scrutiny.

While researching Rose on the Internet, I found other bands hooked to serial killers. There was Marilyn Manson and the Spooky Kids—with band members Berkowitz, Speck, Lucas, and Gacy.

Interestingly, Marilyn Manson's group was under contract to Trent Reznor, of Nine Inch Nails. Reznor had recorded his latest album in the living room of Sharon's former house on Cielo Drive (dubbed Le Pig of Beverly Hills Studio). His CD included the songs *Piggy*, *March of the Pigs*, and *Big Man with a Gun*—reputedly introduced with a barrage of gunshots.

Along with the chart toppers, I found a number of small underground groups who played Manson's music in clubs around the world. Each one of the homage bands pinched at the heart wound of Sharon's murder, but when I learned that Axl Rose planned to release one of Manson's songs as part of his next compilation, it squeezed my heart to pulp.

Manson composed "Look at Your Game Girl" in 1968. Why did Rose want to cover the song twenty-five years later?

"Manson is a dark part of American culture and history," Rose commented to the press. "He's the subject of fear and fascination through books, movies, and the interviews he's done. Most people haven't heard anything Charles Manson recorded. Personally, I liked the lyrics and the melody of the song. Hearing it shocked me, and I thought there might be other people who would like to hear it."

Thanks, but I could have done without it.

My personal appeal to David Geffen to remove the covered song before Guns N' Roses released their upcoming CD, *The Spaghetti Incident?*, was answered with a short press release that he sent from the Caribbean: "I would hope that if Axl Rose had realized how offensive people would find this, he would not have ever recorded this song in the first place. The issue is not the song itself, it's the fact that Charles Manson would be earning money based on the fame he derived committing one of the most horrific crimes of the twentieth century. It's unthinkable to me."

David Geffen had a point, but what lay at the core of my problem with Axl Rose's recording the song and including it on his album was the fact that he was idealizing Charles Manson, putting him on a glamorous pedestal, and sending that message to his fans—and a majority of Guns N' Roses fans are under the age of twenty-one.

Not long before my mother passed away, we went to the market. She stopped the car in the middle of a row of parking spaces and said, "Would you look at that." She pointed to a car parked crookedly near the end of the row. "That car caused a chain-reaction to the entire row; no one can park within the lines. So what happens? Now, the handicapped space is useless because the last car is halfway in it. That's all it takes, one person's inconsideration."

Mom was right, and Axl Rose had the capability of reaching millions.

The most important message that I wanted to instill in my children while they were impressionable was that their actions would always affect another. Like Father O'Reilly said at Sharon's funeral: We create in every act of good we do; we destroy in every act of evil we perform. Axl Rose was about to accentuate that lesson for my kids.

Beginning a national boycott against Geffen Records was a tough call. Ultimately, I was attempting to squelch Axl Rose's freedom of speech. But just as he had a right to express himself, so did I. On December 7, 1993, I launched a boycott against all products released by Geffen, and then hit the talk-show circuit to promote the boycott and demote Manson's current following.

Patti 1994

"Charles Manson is not evil; it's the society that condemns him that's evil."

"Manson didn't kill anyone; he doesn't deserve to be in prison."

"He's a prophet."

"He's the only modern American hero!"

Dissension was the name of the game in a cold Chicago soundstage where *The Bertice Berry Show* was being taped. The moment the cameras began recording, random shouting between guests replaced the usual talk-show format, with the opposing sides more preoccupied with victory than with communication.

With the help of Axl Rose and Richard Lemmons (the manufacturer of the Manson apparel), a fresh breed of believers in Charles Manson's innocence was spawned.

Like many, I held the opinion that *Manson* and *evil* were synonymous. To his proponents, Manson was a peace-loving musician unjustly convicted of murder. How could each side see two completely different personas emanating from the same man?

Seated next to me on the stage for Bertice Berry's show was my sole teammate for the day, Vince Bugliosi. We had been ambushed. Spread throughout the audience were Manson defenders, identifiable by the shirts and hats they wore with his face emblazoned across the fabric. Lemmons sat in the front row.

Sharing the stage with us was our primary opponent, Sandra Good, Manson's most avid and infamous follower since 1969.

When the cameras went to standby for a commercial break, Sandra came over and knelt at my feet. "Patti, you've been lied to for twenty-six years," she said, handing me a business card. "Call me. I will answer any questions you have. Sharon wasn't supposed to be home; *she wasn't supposed to die.* There's so much you don't know about Holly-wood, it's a hellish place to be. And let me tell you, Sharon found out the hard way. If she hadn't decided to get stoned on coke that night, she'd be alive today."

I handed her back the card. "You don't get it, do you, Sandra? My sister did die; the rest of this nonsense you're peddling is incidental."

"You don't understand," she insisted. "The murders had nothing to do with Helter Skelter, that only meant confusion. We saw the country spiraling into a greater state of confusion, which is what you have now."

"And you're making it worse," I told her.

"It can't get much worse than it already is."

"Why murder?" I asked, genuinely trying to understand.

The stage manager interrupted us. The show was seconds from resuming, so the answers to my question would remain a mystery.

The moment the cameras rolled again, the chaos rolled with them. Barely audible through the clamor of the rambunctious crowd, Vince attempted to make a point. "These murders were committed for two reasons: Manson's lust for killing, and his enormous amount of hostil-ity against society."

"Oh, that's bullshit, Vince!" Sandra said. "We killed for the love of brother. We killed to get a brother out of jail! . . . In war . . . you defend yourself."

"That's not self-defense, that's called murder," Vince shot back.

"Oh, yeah?" Sandra laughed. "Why don't we ask Patti about her dad, Colonel Tate. How many lives were lost because of his activities? We were no different!"

I moved to the edge of my seat. "*Do not* compare my father helping to keep this country safe with your group's cold-blooded murders."

She waved me off. "Patti, all I was trying to say is that we were all at war. In the context of the sixties, violence wasn't unusual. Even the music was singing it to us. [With the murders] we showed America violence in their own homes and said, 'This is where our country is headed. *We are your children, doing what you have taught us.*'"

The teenagers in the audience stood, cheering Sandra's perception of them. Many jeered at me when moments later I challenged her. . . . "You know, I'd like to ask, why did my sister have to pay for it—"

"Patti, she shouldn't have been home—" Sandra started.

"Why any of them? They were good people!"

Sandra started to interrupt me again, but I continued. "No! Stop it! The truth is that Manson, Atkins, Krenwinkel, Van Houten, Watson—and in my opinion—Linda Kasabian, are murderers, and we have to protect ourselves from them so they don't go out and kill again. You talk about being at war, and how we watched our soldiers coming home in body bags. Does that justify me watching my sister being rolled out in a body bag? With her belly out to here? . . . He took my sister's God-given right to live."

"He did not kill your sister, and he did not order your sister to be killed, Patti. You've got to believe me."

Thankfully, another commercial break intervened, and I escaped to the restroom, where I rested and prayed to keep my façade of strength from crumbling. I needed the whole bottle, but took just a half of Xanax and slipped it under my tongue, hoping it would take effect before I returned to the stage. When I took to my seat again, Bertice invited four teenagers to join the guests. Steven, Chris, Eric, and Eva all had a common bond: Manson was their role model.

"Why do you follow Charles Manson?" Bertice asked Steven.

"I don't *follow* him. I *identify* with him. He was a white man who had a hard life, who grew up in the same conditions that I did. He's innocent of what you've convicted him of. . . ."

Vince was as dumbfounded at their perception of Manson as I was. He sat forward to address all four of them. "I'd like to ask you all a question for clarification purpose, because there is some ambiguity here. Manson was convicted of nine murders. Now, do you look up to him because you think it's perfectly okay to do what he did, or do you think the convictions are in error?"

His question incited another free-for-all. "Retrial! Retrial!" Richard Lemmons yelled from his seat.

"Show me the proof!". . .

"He didn't kill anyone!". . .

"It's all media hype!". . .

Through the thunder of voices, one lashed out, "You don't get it. Charlie's philosophy is what's important! We identify with it." Steven pointed at me. "We couldn't care less about your sister! This has nothing to do with her."

Though I searched for a reply, the insensitive words that spilled so easily from him left me speechless. A moment later, any thoughts I may have responded with became pointless. The audience rally escalated into a congregated chant of "Free Manson." Momentarily defeated, I felt vulnerable and alone amid this unsettling crowd. Vulnerable, because my childhood fear of confrontations turning violent hadn't dissipated in the slightest. On top of that, last week I received a threatening letter in the mail: *"So sorry to hear about your sis, but when the Manson sisses get out of jail, what they have planned for you will make everyone finally forget about your sis's murder."*

I scanned the studio. Was the author of this note in the audience? On the stage? Sandra? Richard Lemmons?

I withdrew from the uproar, seeking the inner strength to see the show through to the end. With closed eyes, I concentrated on Mom's adage: focus and come out fighting.

"Manson never advocated killing," Steven said. "He never advocated Helter Skelter. He merely recognized the fact that the apocalypse was coming.". . .

Focus and come out fighting.

"He helped people to find their hearts. He told people to do what was in their hearts," Sandra said. . . .

Focus and come out fighting.

"You all keep saying that Manson himself never killed anyone," Vince said, "but in addition to the seven Tate/LaBianca murders he was also convicted of physically participating in the murders of Shorty Shea and Gary Hinman. . . ."

"There's no proof," Christina shouted.

"No proof?" Vince laughed. "A jury thought there was enough proof and convicted him.". . .

Focus and come out fighting.

"Hold on a minute everyone," Bertice said. "I need to introduce Richard Lemmons. Richard, why do you make T-shirts with Charles Manson's image on them?"

"I feel that Charlie got a raw deal. . . . And I want to say that I'm so tired of hearing Patricia Tate whine about how much money we're making. Why don't you ask Bugliosi how many millions he's made off his book?"

Focus and come out fighting.

"Listen," Vince answered. "It's perfectly acceptable for someone to write a nonfiction book about their experiences—"

Lemmons yelled, "It's fiction—"

"—whether it's a lawyer in a case, a professional athlete, or a general during wartime. That's all I did. Don't equate that with Manson's being responsible for nine murders, and profiting from the murders he committed through you! You're glorifying him. On that same note, it's a sad commentary on justice in America that a murderer who was supposed to receive the death penalty ends up having his song appear on a hit rock album by Guns N' Roses. From a moral standpoint, it's truly distasteful."

"Hey! People are so ticked off at Charlie; where's Tex Watson come up in all this?" Lemmons stood and turned toward the audience. "Tex Watson did eighty-five percent of the killings. Bunch of ignorant people, you don't even know who Tex is!". . .

Focus and come out fighting.

"Patti," Bertice asked, "you've been very quiet, but you see all of this happening. What would you like to say?"

"What I'd like to say to Mr. Lemmons—and I've already spoken to you about this in depth—do you see this madness? You're adding to it."

"I ain't adding to nothing!"

Focus. "What I'd like to say to the young people who reflect on Manson's beliefs—find something positive to reflect on. Get out there and love one another. I grew up hard with the evil that surrounded me with my sister's death. It was the destruction of my world. Please don't reflect on that evil. Get out there and rebel in a positive way."

The show ended, and as was my custom, I wasted no time getting out of the studio and into the waiting car. The difference this time was that I was sure that Sandra Good was following, and I was scared.

When I decided to see through my commitment to Mom, I jumped in with both feet—like a cat thrown into icy water. Some days, I thrashed about, terrified, accomplishing little more than a lot of chaotic splashing. Some days, I pulled myself out, shook off the frigid water, and focused on what's at stake. And it's not only my family's justice on the line. Because the Manson Family are considered to be the most heinous killers in the annals of California history, releasing any one of them will no doubt precipitate the release of countless other heinous killers in the California prison system.

"Long day?" the driver asked.

"And it's only just begun," I sighed.

"I caught most of the show from backstage. I've got to hand it to you—if it'd been me up there, I would have slugged that Lemmons guy, but you held your cool, you did good."

"I don't think I did much good in there today."

"Hey, for what it's worth, I'm going straight home and going through my kid's music. It just takes one bad apple to ruin them. You know what I'm saying?" He started the car. "Are we still going straight to the airport?"

"Yes. Thank you," I said, and bundled up against the snow scuttling around the car. As we pulled away, I took one last look to be sure that Sandra wasn't following.

I closed my eyes and tried to rest before round two. With less than eight hours to spare, I had a meeting with Geffen Records in Los Angeles.

THE WEATHER IN West Hollywood wasn't much better than what I'd left behind in Chicago. The rain hammered against the window of the conference room where I waited for Geffen executive Ed Rosenblatt.

Jet-lagged and jittery from the NoDoz I'd taken, I didn't have very high expectations for the meeting. Geffen had released the Guns N' Roses CD with the Manson song included. Overseas sales had sold an estimated three hundred thousand copies. The most I hoped for was to touch them with my sister's story and my love for her, so it might deter them from doing anything like this again in the future.

The door opened. Six men entered. Five of them scattered to seats around the table. One took the helm. "Sorry to keep you waiting," Rosenblatt announced. "Please, sit down. Can I get you some coffee?"

"I'm fine," I said, wondering who his silent partners were, positive that at least one of them had a law degree. I wasted no time and held up a photo of Sharon. "On August eighth, 1969, my sister woke up without a care in the world. She had an early appointment with her obstetrician, and called later to tell us that the baby was fine. Around one o'clock, Sharon had lunch with friends. From three to five, she napped. Later, she went to dinner with friends. When she got home, she retired to her bedroom with plans on waking to another day. What happened after that is indescribable.

"I look at her picture, and all of the emotions from twenty-five years ago come back to me as if it were yesterday. What I remember most about the time after Sharon was murdered, is watching my parents suffer. Knowing I could never fill Sharon's shoes, and realizing that all I could do was wrap my arms around my mom and dad, tell them how much I love them, and go on from day to uncertain day.

"In light of that experience, I have only three goals in my life right now. The first is to raise my kids in an environment of peace and security. Second, to see to it that Sharon's killers remain in prison. And the third, to make every attempt to have less victimization in our country. As I see it, your company is an obstacle for all three of my goals."

Rosenblatt shifted in his chair. "I'm sure that you can appreciate that our company would have preferred that the Manson song wasn't on the Guns N' Roses album, but you must understand, that the choice to cover Mr. Manson's song was the sole decision of Axl Rose and in no way represents the opinion of Geffen Records."

"You could have chosen not to publish the song," I said.

"That's not true. Under our legal agreement with the band, they have complete creative rights to include that song on their album."

"Are you saying that if Axl Rose wanted to put a repulsively offending song on his album that you couldn't stop him?"

"Quite frankly, Miss Tate, there's nothing even remotely repulsive about Mr. Manson's song—"

"That's a matter of opinion," I interjected.

"Yes, well, given our belief in freedom of speech, that's a subjective consideration. If people are offended, they don't have to buy the album." He sifted through his briefcase. "Clearly, we are at an impasse. But to show our good faith, sensitivity, and support to victims of crime everywhere, we would like to present you with this check as our contribution to the Doris Tate Crime Victims Bureau." He put his offer on the table and slid it toward me.

I put Sharon's death photo on the table and slid it toward him. "This is what Manson is about, and by publishing his song, this is what you're promoting."

"Miss Tate, I find your business tactics distasteful."

"Then I guess we're even. Keep your blood money."

Rosenblatt stood and the others followed his lead. "This meeting is over. Under the circumstances, Geffen will decline any future meetings with you or your organization."

The entourage left in unison. The check remained. I picked it up, shredded it into thirty-five pieces and left them atop a note: "Mr. Rosenblatt, Charles Manson may be responsible for as many as thirty-five deaths. No amount of money will hush their cry for justice."

20

THE SHOW MUST GO ON

Thirty years later, there's hardly a day that I don't think about Sharon. She's still in my heart and will be until the day I die. I miss her. It's like learning to live with a gaping wound that never heals. . . .

—PATTI TATE

Patti

"You know Sharon Tate, the victim," the television announcer declared. "Now meet Sharon Tate, the person. This is her story, the E! *True Hollywood Story.* Sharon Tate's fantastic life and tragic death will be told through photos and incredible, never-before-seen archival footage. You'll hear from friends, costars, investigators, prosecutors, and for the first time ever on television, you'll hear the terrifying account of William Garretson, the lone survivor in a night of mayhem."

One person E! viewers would not hear from was me; a day before my scheduled interview for the show, I was forced to cancel the appointment.

ALMOST A YEAR ago, I felt a lump on the side of my right breast. After days of agonized waiting for the test results, I sat in the doctor's office, imagining that moment of relief when he said, "The tumor is benign." Instead, I heard the word *cancer*, and after that, everything else hazed over my mind. My head throbbed to the beat of terror and a single thought—I'm going to die just like Mom. Spooked by her cancerous outcome, I saw myself as a patient. I saw myself lying in a coma. I saw myself dead.

I never thought to ask how or why I got cancer. According to the expert's statistics, I had turned my body into a malignancy welcome wagon. I was a twenty-year smoker. Everything I ate was unhealthy. I had color-treated my hair so many times that I should have gotten a brain tumor. I drank a couple of beers or glasses of wine each night. And the front-runner: my bouts of stress, anxiety, and depression.

Beyond a doubt, the doctor said, the tumor had started growing at least five years ago, placing the onset at the end of my mother's life. And Lord knows those were impossible times that left a gaping wound that never healed. I still missed my mother as strongly as the moment she passed and needed her now more than ever.

Within a month of my diagnosis, I had a lumpectomy, a Port-A-Cath inserted in my chest, and the first of nine chemotherapy treatments.

Loved ones pulled me in twenty directions to counter the deadly cells: eat organic, take vitamins, see an acupuncturist, take the holistic route, learn how to meditate, do yoga twice a day, write in a journal, exercise, get counseling.

I appreciated all their well-meaning advice. The thing was, I didn't want to change my ways; I just wanted my life back.

My stubbornness prevailed for a while because I wasn't afraid of dying. I was, however, petrified to leave my kids motherless. So, like it or not, I changed my lifestyle and followed everyone's strategy on how to rid the cancer.

I cleared my mind of comfort foods like Mom's pancakes smothered in butter and syrup, grilled cheese, pizza, milkshakes, coffee, and chocolate, then filled the refrigerator with tofu and organic produce. Soy energy shakes replaced the milk shakes, and I pitched the coffee for a cancer tea that no amount of honey and lemon could disguise.

I swallowed vitamins until there wasn't room in my belly for food. I learned who loved me unconditionally, because they were the only ones kissing me after all the cancer-hating garlic and shark cartilage I consumed. I learned breathing exercises (cancer hates oxygen), and visualization exercises (healthy cells bombing cancer cells). I drank so much water to flush the toxins that I went through a roll of toilet paper a day. And I took walks on the beach when I barely had the energy to get to the bathroom. But it was all worth it, even if the end result was just one more minute with my kids.

I read a book by Dr. Bernie Siegel filled with love, filled with medicine, and filled with miracles. I wanted to be one of the miracles. Halfway through the book, I found my biography on one of his pages: "One of the most common precursors of cancer is a traumatic loss. . . . Typical depressed patients, by abdicating normal activity, are at least offering some response to what they perceive as an unbearable situation. . . . They continue with their routines and an outward show of happiness, when on the inside their lives have come to lack all meaning. Their state is quiet desperation, meek and obliging on the outside, but filled with unacknowledged rage and frustration."

If I could find the diagnosis, surely, I could find the cure; and so began my journey into the world of psychology.

Go see seven different psychologists and you'll get ten different opinions about the unaligned gears grinding inside.

The first doctor seemed more interested in finding out if, as a child, I'd ever had sex with Roman—suggesting that there was a chance that I had repressed sexual hang-ups because of the experience—hence the attack on my breast. No.

Another doctor grilled me about the parties I went to as a kid. "Which movie stars did you meet?" Next.

"Maybe your father didn't love you enough as a little girl, and this is your way of getting his attention." That didn't ring true in the slightest. Next.

"As a child, you were jealous of Sharon's stardom. Your cancer is a subconscious scream for the attention she stole from you." Never.

I left the last doctor's office wondering which of us needed therapy more.

Disenchanted as I was, each of the psychologists added to a common thread that merited my attention. I had survivor's guilt and because of that guilt, I didn't feel worthy of living. I couldn't deny the truth of that idea. It first came to me on the day of Sharon's funeral. Why had God forsaken Sharon, yet deemed me fit to continue? The question haunted me into adulthood; Mom's passing reinforced it because I was sure that they both had so much more to offer the world than I did.

The other commonality the psychologists chased was the notion that I had to release my hostility toward Sharon's killers because cancer cells thrive on anger.

For months, I struggled to forgive the killers. I prayed for them. I visualized hugging them—oy. I spoke the words: "Susan Atkins, I forgive you and have love for you." I wrote it in my journal, and mapped out a triangle of anger and forgiveness.

Each night before bed and every morning before I got up, I prayed for God's support of mercy and love for those I'd hated for so long.

I tried and tried and tried to whitewash my disdain for Sharon's killers, but I couldn't. Ultimately, I felt I was making a deal with the devil—give me another year and I'll forgive them, give me two, and I'll love them. The entire concept felt shameful. If that's what it took to save myself, to hell with it. I gave up on the quest and booked a first-class ticket on the next flight to Satanville.

Often, I looked in the mirror to try to find the beautiful girl in the Saks Fifth Avenue store, standing next to Sharon, making a promise. Instead, the mirror reflected an unwelcome stranger. My hair fell out five chemo treatments ago, my eyebrows and lashes disappeared two treatments ago. The steroids they injected me with bloated my face until my eyes seemed to retreat into their sockets. I tripped into another sinkhole; I couldn't even keep my promise to Sharon.

By the time I found the Wellness Community, I was sitting on the edge of my grave, legs dangling, ready to take the plunge. I'd heard about the cancer support group months ago, but I didn't believe that meeting with other sick people, talking about sickness, would be helpful or healing.

During my second-to-last chemo treatment, in walked Betsy. She was tall and beautiful despite her baldness, and filled with refreshing optimism.

Set comfortably in baby-blue Barcaloungers, we struck up the usual chemo room banter: How did you find it? Surgery? Radiation? What kind of chemo? How many? Blood count? Tumor markers? She stumped me with the last one: "What are you going to do when this is over?"

I didn't have a clue. For six months, I had been riding a merry-go-round of treatments. The future seemed illusory.

With dreadful anticipation I watched the nurse flush my Port-A-Cath; next came the antinausea medication, steroids, then the chemo. Where *would* my life go after this routine?

The office counselor came in while the chemo slowly drip-dripped into my vein. She handed Betsy and me a flyer promoting Joke Fest Night at the Wellness Community. They have a joke fest? I like to laugh—I love to laugh, maybe I *should* give it a try. Then she handed us another bulletin for an upcoming group session for newly diagnosed cancer patients. Betsy nudged me. "If you go, I'll go."

Four weeks later, I sat in a room surrounded by other cancer patients, each as frightened and uncertain as I was feeling. It instantly reminded me of Mom's first Parents of Murdered Children meeting; cancer is no less biased than murder is to its victims.

We went around the room sharing our cancer experiences, each entirely different, yet completely the same.

When it was my turn, I drenched a handkerchief while telling my story. My tears were long overdue and therapeutically cleansing. "I wake up from night terrors, hearing my kids calling for me. I don't want to leave them," I said.

"Then make plans not to leave them," the counselor advised. "Instead of expelling all that negative energy about what they'll do without you, surround yourself with positive plans for the future. See yourself at their graduations. See yourself at their weddings. Imagine rocking your grandchildren."

Norma was the first to analyze my other hurdle on the healing path. "Seems to me that in the process of trying to absolve these killers, you've crucified yourself."

"Yeah," Janice said, "you've turned them into the good guys, and made yourself the villain."

"Personally, I think this forgiveness stuff is overrated," Mindy said. "When my husband left me for another woman, I fantasized about taking a bat to his balls—and it felt damned good! My point is, God says it's okay to have those feelings; you just can't act on them. In fact, you might try doing some visualization exercises of the killers. Imagine running over them with your car a couple of times—now that's healing!" she laughed.

We all laughed, and God, did it feel good.

Our group leader stood. "We're here to help you learn the tools for the pursuit of just this type of happiness so you can be an active participant instead of a hopeless, helpless, passive victim."

The last word caught my attention and triggered a new fight. I did not want to be another type of victim—one was bad enough. It was time to focus and come out fighting.

As the weeks passed at the Wellness Community, I found a group of six soul mates within our group. Those six believed in love, medicine, and miracles. They believed in healing. They celebrated life. They made me brave. They nourished my soul. They were my Bosom Buddies.

Our group was diverse. Deb could make me laugh till I wet my pants. Barbara, the den mother. Bonnie, with big, beautiful eyes and a heart to match. Betsy—young, vibrant, and hope-filled. In Shirley I saw my mother's strength. With Cindy, I saw myself: shy and quiet.

The Bosom Buddies shared potluck dinners. We shared our lives, our fears, our joys, and together, we laughed more than seven cancer patients seem to have a right to. Equally important, the BB's were unfazed by my notoriety and kept my cancer a secret. My illness was not for public consumption. I didn't want pity. I didn't want Sharon's killers to see me weak. When the media called, I declined with lame excuses to hide the truth.

I FLUFFED THE pillows on my bed as E's *The Last Days of Sharon Tate* began. Perhaps if I'd been honest with the show's producers, they would have been more sympathetic to my family. As it was, they interviewed biographer Greg King, who unfairly portrayed my father as a frostbitten tyrant and Mom as a starstruck, meddling, conniving stage mother, pushing Sharon into the limelight.

On the opposite side of the coin, I adored the program's archival film clips of Sharon, many of which I'd never seen.

The first hour was filled with interviews from Sharon's high school buddies as well as friends I remembered from the later 1960s. Sharon's first agent, Hal Gefsky, appeared as sweet as ever while he told his story of discovering Sharon. And Sharon's friend Sheilah Wells, talked about their careers, parties, Jay, and Roman.

They geared the second hour toward Manson and the murders. Ordinarily at that point, I would have turned off the television. This time I waited to see the hyped interview with William Garretson. Mom always believed that he'd lied in 1969 when he claimed he didn't see or hear anything the night of the murders. His thirty-year-old, revised memory would be a curious revelation. There was another more intriguing reason—Garretson was engaged to "Sharon's daughter."

"Conspiracy buffs have a way to go to top this one on Sharon Tate," columnist Steve Stephens wrote in the *Columbus Dispatch*. "For thirty years, William Garretson refused to speak about the Manson Family murders. Now, he finds himself at the center of a conspiracy theory that gives 'Elvis lives' a run for its money.

"Imagine that the unborn child of director Roman Polanski and actress Sharon Tate didn't die with Tate in the Manson Family attack. Instead, the baby was delivered by celebrity hairdresser Jay Sebring— using his barber tools—just before he and Tate died. Then, Frank

Sinatra, with help from the Genovese crime family, spirited the child to New Jersey on a private jet. Next, Patty Duke's cousin raised the baby to protect it from Manson retribution. And finally, the child grew up to fall in love with Garretson.

"That is the story of Rosie Blanchard, aka Rosie Tate-Polanski."

Ridiculous as it was, the article wasn't startling. Rosie's name first came to light when she initiated contact with my parents a couple of years ago. At that time, she had a different story—she was Sharon reincarnated. Instead of coming right out and saying she was Sharon, she'd send them mysterious cards on special occasions such as Mother's Day. Eventually, she showed up at my parents' house on the anniversary of Sharon's death. I wasn't there at the time, but Dad said she got as far as "Hi, I'm your daughter, Sharon" before he slammed the door in her face.

After Mom passed away, Rosie turned her attention toward me. I certainly didn't have proof one way or the other about whether she could really be Sharon reincarnated, nor could I predict what her reaction would be if I resisted her mind trip. So I played along and adhered to my rule of always keeping potential enemies close at hand where you can keep an eye on them.

Over the phone and miles between us Rosie's delusion seemed harmless, though she unnerved me when she showed up in person at my first parole hearing for Susan Atkins. Call it sixth sense, call it before-the-hearing jitters—my instincts told me there was something inherently wrong with Rosie, particularly because I'd asked her to stay away from the proceedings that morning.

After the hearing, I kept her at a distance, busying myself with reporters. Still, I kept a watchful eye, wondering if a dormant bomb was about to explode.

I let the charade play out until my cancer diagnosis. At that point, I wanted to tie up some loose ends—just in case it was my time to go see the good Lord. Rosie was not only a loose end, but a fraying one that could rip at any moment. After I was gone, I didn't want my kids to pick up the phone and hear, "Hi, it's Aunt Sharon."

Rosie's apartment was set in the heart of Burbank. It had to be over a hundred degrees the day I knocked on her door. Inside, a noisy box fan, as useless as a Popsicle stored in an empty ice chest, stirred the overperfumed air. She gave me the grand tour, which, considering the two-room total, was quick.

The bedroom decor consisted of a mattress on the floor and moving boxes. In the living room stood a breakfast table, two folding lawn chairs, and more boxes. "Sorry for the mess," she said from the kitchen area. "I'm going back East for a few months to regroup. There was a lot more furniture in here, but the movers took it yesterday."

I didn't believe her. She continued packing, avoiding eye contact. "Rosie, I'm going to get right to the point of my visit. I don't believe that you are Sharon reincarnated."

Her back was toward me. She froze midmotion. Seconds passed before she slammed a saucer into a packing box. Quiet again. Then she twirled around with a grin that was much too big. "I'm glad you said it, because there's something I've been wanting to tell you, but I haven't had the nerve—I'm Sharon's baby."

"Wha—"

"I'm your niece! I've been alive all this time!" she said, starting toward me.

A bead of sweat trickled down my back, chilling me on its way. "Rosie, stop it. Sharon's baby died with her. It was a boy, and he's buried with her."

The smile disappeared. "Well, for once you're wrong, Miss Smarty Pants," she said, turning back to the task of packing. "Sharon *is* my mother, and that's all there is to it." She slammed another dish into the box.

"You're confused or mistaken or—"

Hostility flared in her nostrils, her lips curled. "If you don't believe me, I want you out of my apartment!"

A butcher knife was within her reach. I edged toward the door, feeling like a caged rat. "Okay, I'll leave, but I don't want you to contact anyone in my family again. I'm having my number changed—"

"Get out, get out, get out!"

The severed relationship lasted about two months, and then a fax came through with her real intention. The first page was the 1969 court order transfer of Sharon's estate from Roman to Dad. The pages following summarized her claim that since Rosie was Sharon's daughter, she was part of Sharon's estate; an estate that Dad had inherited. Since Rosie was, to her mind, Roman's child, she believed she was entitled to his childcare and given name. The fragmented document ran another four pages, but the long story made short is that she wanted money.

Somewhere along the line, Rosie ended up in Ohio, with Garretson. In the course of their relationship, she convinced him that he'd seen her in 1969 at the crime scene before they whisked her away. "Someone should take this young lady seriously," Garretson said of Rosie. "There's a strong possibility that she is Sharon Tate's baby."

The rest of Garretson's testimonial on the E! channel was equally troublesome. "I lied to the police [in 1969] and told them that I'd been listening to music all night and heard nothing," Garretson said. "I was in a state of shock. They wanted answers that I couldn't give them. *I did* hear a scream that night. The scream sounded like someone was about to be thrown in the pool or something. I looked through the window and there was a girl chasing a girl. I wondered what was going on, but I didn't want to look like I was spying from the window. Then I heard someone screaming, 'Stop, stop. I'm already dead.' After that, I closed the window and went to sleep."

The odd thing about his memory is that the quote, "Stop, I'm already dead," came from Patricia Krenwinkel's parole hearing—she was quoting Leno LaBianca screaming "Stop stabbing me, I'm already dead." I hope that Mr. Garretson's fresh recollection is merely an attempt to catch his fifteen minutes of fame, for that is much more acceptable an act than closing a window and ignoring screams for help.

If I'd been part of the show . . . I stopped midthought, frustrated by my forced silence, and finally understanding Mom's insistence on doing every interview that came her way. Had I done the show, no one would have heard Greg King's interpretation of my family. I could have balanced fact and fiction.

The show had rustled old memories like fallen leaves on a windy day, memories that brought first joy, and then another bout of depression as I fretted over the possibility of my cancer terminally silencing my opinion.

21

AND THEN SOME

If someone participates in a particularly heinous, nightmarish murder
the way that Leslie Van Houten did, it seems to me that there must
be something in the deepest recesses of that person's soul that
enables them to do what they did. Something that fortunately you
and I do not have, because we would never do what she did. And
therefore, it seems to me, that to release this type of person on a
vulnerable society is to take a risk that we probably shouldn't take.

VINCENT BUGLIOSI

Brie

I watched the *Larry King Show*, impatient to see my mother's segment. The precursor to the live discussion was an interview between
King and Leslie Van Houten. What disturbed me was the demure,
straightforward sincerity with which Van Houten answered King's
questions. I liked her. And a likable killer is a dangerous combination
that could spring the confining lock.

Van Houten is a pawn to my family. With just two murder convictions under her belt (for Mr. and Mrs. LaBianca), her chances for parole are the greatest of all the Manson Family; therefore, she is a pawn
with a strategic position on the board. If she walked through society's

pardoning doors, they were unlikely to close until Watson, Atkins, Krenwinkel, and even Manson followed her trail.

Since Mama couldn't attend Van Houten's parole hearing, she indirectly sent her message to the board in a venue that was open to her—television. Leslie's father, Paul Van Houten, didn't take kindly to Mama's venture. They were both panelists on Larry King's show, along with Vincent Bugliosi.

"Paul, this had to be an incredible shock to you when this arrest occurred. Can you remember back to that time?" King asked.

"I'm not interested. That's history. I've never asked Leslie about the murders, we've never talked about the murders. I have not let this case run my life. I've accepted it, I live my life. I don't go to bed every night thinking about it."

"Patti, how do you feel about Leslie?" King asked.

"The state commuted her death sentence to life in prison, that's her second chance, not freedom. As far as rehabilitation, that shouldn't be a consideration for her or any first-degree murderer. You take a life; you pay for it the rest of your life, in prison."

"Now, Paul, what would you say to Patti?" King asked.

"Leslie had nothing to do with Patti Tate's sister. If the LaBiancas want to talk about it, they have the right.". . .

"Paul, was there anything in Leslie's childhood that could have told you this was coming?"

"No. If Leslie had never smoked her first marijuana cigarette, this never would have happened."

"You're blaming it on marijuana?" King asked.

"Marijuana put her with the group and Manson was able to maneuver these people with drugs."

"But millions have smoked marijuana and didn't go kill people," King said.

At a later point in the interview, Bugliosi would emphasize Van Houten's accountability when he said, "Larry, during the penalty trial, psychologist Joel Hochman examined Leslie. She told him how much she loved the world and the people inside of it, whereupon he said to her, 'Leslie, if you love people, how could you do what you did?' And she responded, 'Well, that's something inside me, too.' At the time, I sensed that of the three female defendants, Leslie was the least committed to Manson. And lo and behold, Dr. Hochman came to the same opinion. That increases her personal responsibility and her moral culpability."

Toward the end of the show, King said, "Patti, you must have some sympathy for Paul and his daughter."

"I have sympathy for her victim's families. The ones that are really left with nothing, nothing but memories. Mr. Van Houten has his daughter; she has her life. I think that's more than enough."

I pressed the PAUSE button and stared longingly at my mother on the screen.

I'd tightly drawn the shades to block the daylight glare in the family room. It was the same family room in which my mother had scrambled her way out of adolescence. The same family room where Nana had hugged me, smothering my tiny body against her ample bosom, where it was smooshy and safe. The same family room where Papa cradled me with one hand and stoked his cigar with the other. The same family room where I will soon cuddle with my first baby.

They're all gone now, and daily, as I ramble through our home, I'm taken back in time to hear the voices of my past. Mama's ritual morning call, "Briezzy, get up. You're going to be late for school." Or her quiet "boo" against my ear as she snuck up behind me. Or Nana's high-pitched Texas twang announcing our visits, "Comp-any!"

Though I can't free-float in the pool as well as Nana, or look as dashing as she did wearing a bowl over her head to make oatmeal enticing, I have inherited many of her qualities. Some are physical, like her pudgy nose, thinning hair, and bladder infections. Others I acquired, like her enthusiasm for Christmas. She decorated the house with the glitz of a Vegas casino and the warmth of a jigger of Southern Comfort; all surrounding a towering tree stacked knee-high with gifts. Along the line, I picked up her penchant to cuss; not the nasty words, but *shit* and *hell* were her favorites—usually followed by "P.J." as she scrapped with Papa.

My grandparents had a unique relationship—they constantly bickered, but that never shadowed the love and commitment of their fifty years of marriage. They were polar opposites. She loved unconditionally; he suspected everyone. She countered his inhibited nature with exuberant nurturing. Both were as set in their ways as a grape stain to white pants and equally as stubborn. Despite it all, their love was as preserved and age-worn as a pressed rose hidden in a Bible. Happily, I have inherited the best qualities of both of them.

Too cantankerous to find it himself, new love instead found Papa after Nana died. When he moved out of our house to live with his girlfriend, he failed to take the aroma of his constantly burning pipe or cigar tobacco. I despised the smoky atmosphere when he lived here. In his absence, I sniff the upholstery to bring him home.

In the 1950s Papa went AWOL twice from the army. Once to see Sharon right after birth, the second to bring Nana home after she left him (the reason for which neither of them could—or chose to—remember). When he died of congestive heart failure in 2005, I believe he went AWOL from life to recapture all his girls and heal his heart.

Nana's bedroom is now mine, and though it's changed, I can still close my eyes to see it as it was during her lifetime. Soft flower patterns covered the curtains and bed linens; family pictures fought for prominence in the rest of the room, including the surface of her nightstands, which she stuffed with caramels and mint chocolates.

Nana didn't sleep with underwear, but she did sleep with a gun—either of which, if flashed, she assured me, would scare away any intruder. She wasn't the only one to sleep with a gun—so did Papa, and so did Mama. On my tenth birthday, I discovered their need for security.

Standing tiptoe atop a chair, I rummaged through my mother's dresser drawer looking for a scarf. What I found instead was a plastic baggie filled with pictures. I recognized the top photo. It was Aunt Sharon's house. I glanced over my shoulder to be sure I was still alone. The next picture was Sharon, curled up on the floor, distorted by blood, unrecognizable except for her pregnant belly. The third picture was more unnerving than the last. Sharon on a morgue table, hair pulled back to unveil her face, eerily alive with open eyes, and mouth curved to a smile as if her last instant of life had revealed something to cause joy. Her face possessed a ghostly resemblance to Mama.

"What are you doing?"

I dropped the forbidden fruit and whirled to find the Boy (Papa's nickname for Bryce, the sole male born into his family in two generations). "Nothing," I said, slamming the drawer closed.

My brother looked at me curiously. "Why are you crying?"

I wiped at my eyes. "I'm not."

"Yes you are. What did you drop into the drawer?"

"Pictures of Aunt Sharon."

"I wanna see." He started toward the dresser.

I grabbed his arm. "No. Let's go."

Bryce was a smart kid for seven. He studied me, deducing the situation with squinted eyes. "They're the bad pictures, aren't they?"

"Shut up," I said, closing us into the hallway.

"I'm telling Mama. You are in so much trouble."

"No shit." I slammed my bedroom door in his face. I was in trouble, but from a self-induced punishment.

Tears splashed on my pillow. For as long as my memory allows, Mama and Nana made Aunt Sharon an active part of my life. She's never dead in my mind, never a victim, never a movie star. She is simply someone who is very real to me, someone I admire, and someone I love. It broke my heart to see a picture that preserved her in such an appalling representation.

Uncovering those police photos was a watershed in my childhood. From that point forward, I knew what a murder victim looked like; therefore, I understood why adults slept near guns. After viewing the pictures, I sympathized with my grandparents' and my mother's loss of Sharon, even felt their pain, even felt their terror that such horrible acts were committed, not by werewolves, vampires, or zombies, but by human beings. The pictures clarified Nana's refusal to look at them; she had the age and wisdom that I lacked. Most of all, the pictures gave reason to Mama's smothering overprotectiveness and inexplicable, saddened pain.

I've only seen Mama break down on a couple of occasions. The first time, she was driving us home from my grandparents' house. I was sandwiched between my brother and sister in the backseat, bickering, oblivious that it was August 9. I can't remember what the argument was about, but I'll never forget Mama's reaction. She pulled the car to the curb. When she turned, it was shocking to see her with gushing tears. "Stop it! My sister is dead! She's gone! Isn't that enough? Appreciate that you have each other right now because it may not last."

The second instance came the day she attended her first parole hearing. Uncontrollable tears ran all morning. Before she left, she hugged me as if she'd never see me again.

A knock on my door hushed my tears. I hesitated to answer, scared of the punishment if it was my mother. The morgue photo flashed again. I didn't care what I had coming. Smiling or frowning, I needed to see Mom's face. "Come in," I managed.

"Briezzy, what's the matter, hon?" she said sliding my head onto her lap.

Assured by her touch, I locked my arms around her waist. "I saw the pictures in your dresser."

I felt her sag with disappointment. "Oh, Briezzy, why did you do that?"

"I didn't mean to. I was looking for a scarf. There's one picture that looks just like you." The morgue photo slammed another punch at my gut. I double-checked that it was Mama's hug shushing the slugger. I expected to see anger, but she looked as sad as I felt. I buried my face in her lap. "I'm sorry."

"So am I, baby," she said, as she ran her soft, cool fingertips over my forehead, down my cheeks, and through my hair until my tears dried.

A thief in the night, cancer stole that unconditional loving solace that radiated from my mother and brightened my life.

Long past midnight, I went to sleep the same as I had for the past twelve months, uncertain of the future. After a two-year remission,

Mama's cancer returned as ravenous as a cougar returning to finish off its hidden prey. The cancer spread into her lungs as quickly as cotton candy builds on a stick. I watched it suck the life from her; I felt it suck the life from me. Still, I was young enough to believe she'd live forever, and she reinforced that faith by being so damned optimistic.

In my presence, Mama veiled her depressive fear behind a smile. A trachea tube attached to a breathing machine paralyzed her melodic voice, but it could not silence her smile. It said I love you. It told me I was her world. It lied with reassurance. In hindsight, it said *I'll miss you.*

In her last days, she fell into a comalike state, completely unresponsive save one gesture. "Mama, can you give me a kiss?" I'd ask, with my lips resting on hers. Without fail, they ever so slightly puckered.

At 5:53 A.M. on June 3, 2000, my bedroom door opened. Mama's friend Lisa leaned against the threshold, silhouetted by the hall light. "Brie, wake up, hon. Your mom's gone."

I sat up, rubbing the sleep from my eyes. "What?"

"Mama's gone."

I dashed past her to Mama's bedroom, positive that she was mistaken.

A single lamp glowed on the nightstand. The breathing machine whished then whooshed without benefit. I glanced around the room at all the angel figurines and pictures that surrounded Mama and knew she'd left me to be with them. Neither cancer nor death had stolen her riveting beauty. Once unplugged, the ventilator also took its final breath and fell quiet.

I got into bed with her for one last snuggle. This time when I rested my lips to hers, I felt a devastatingly still response. Regret is a waste of energy, yet its voice overcame mine. *I should have spent more time with you.* I'm sure that my brother and sister joined me, but I didn't see them, my eyes locked to Mama's face, unable to say good-bye.

The mortuary attendants were quick to arrive—too quick for my needs. The two men wore appropriately black suits. They were courteous and quietly respectful; nevertheless, they irritated the hell out of me. From the white van they arrived in, to the gloves they snapped on, down to the rattling stretcher they wheeled to the bedroom, they were the enemy who planned to steal my mother.

Safely transporting her to a stretcher, the black suits then followed the ritual of lifting the sheet to cover her face. "Stop!" I yelled. They must have thought I was a loon, but I remembered Mama trembling in the hospital ICU each time a patient died. Orderlies wheeled the newly deceased past her room on puce, square gurneys. At the sight of the gurney's coordinating cover, stamped "morgue," she would say, "Don't let them take me out in one of those dead beds, okay? Just a bed."

I grabbed his black sleeve. "Fold the sheet at her chest, like you're tucking her into bed."

The stretcher legs buckled as the black suits slid it into the van. Before Mama vanished from sight, I looked at the men of darkness and echoed her words of wisdom to anyone taking me for a drive, "Be careful. You've got precious cargo in there." I leaned in for a kiss good-bye while repeating her parting comment to my friends and me as we'd back down the driveway, "Have fun, but be careful. I love you tons."

We managed to keep my mother's passing from the press for three days. The day of her funeral, CNN broke the story: "Patricia Tate, sister of slain actress Sharon Tate, victims' rights advocate, Manson Family opponent, loses her battle with breast cancer." But there are significantly more important attributes of my mother that are worth sharing.

On the surface, Patti Tate seemed an uncomplicated person. If asked, everyone she knew would have said the same thing about her: sweet, naïve, and unconditionally loving.

Yet my mother was a complicated woman; below the surface, she was a restless soul who felt unworthy of any of those positive affirmations. In clinical circles, they told her she had survivor's guilt. From my perspective, she was just a little closer to heaven than the rest of us; stuck in a world that had bombarded her with negativity, and in turn, she continually searched for assurance, that somewhere, somehow, love could conquer all.

I'm always astounded when I hear the blame game from criminals who use their tough childhoods as an excuse for their actions. My mom is proof that adverse conditions don't ever need to lead to violence.

She once played a game called Scruples; a game that questions morals. One question haunted her: Would you rip the wing off a butterfly for a million dollars? Her instantaneous answer was no. The question haunted her not because she ever faltered in her response, but because some playing the game with her debated the integrity of her answer. Those who doubted my mother that night did not know her; for question and answer are the essence of who she truly was.

The few times that I saw her raging angry (aside from when she caught me sneaking out one night) were always rooted in injustice or cruelty to another; no matter how minimal those two actions were, they were intolerable. Even if I pinched my brother during a fight, she'd verbally lash out at me, "You don't ever have the right to hurt another being."

While I was in high school, none of my friends called her Ms. Tate or Patti; she was Mama to all of us. She listened to our opinions, and she respected them as if we were adults with something important to contribute to the world.

She was the mama we all went to with our boyfriend problems, and who regarded them as the monumental loves we all thought they were.

Over the weekends, everyone camped out at my house where my mom helped us decide what outfits to wear to the parties, then drove us to those parties, and trusted us to stay even if she saw someone with a beer. And when we did find trouble, she was the one who bailed us out of tight corners, followed by stern lectures while we sulked on the couch.

She was the mama who taught us to say "I love you" at every opportunity. Most important, she was the mama who raised us to be the free spirit she was.

Long before John Edward and Sylvia Browne made it vogue, Holy Cross Cemetery had been a familiar place where I would always go to get a reassuring hug from the other side. There, the headstone of Sharon's grave had been altered twice since she passed, once for Nana, then again for Mama. Below Sharon and Nana's epitaphs is Mama's: OUR MOTHER, OUR INSPIRATION.

Rather than a depository of remains, I like to think of the thrice-used burial plot as a playful gathering of souls.

At Mama's graveside service, we placed her angel urn next to Nana's at the foot of Sharon's casket and released her to them. Then we released butterflies in celebration of her new existence. A life where I'm confident she is as equally excited by being reunited with Nana and Aunt Sharon as she was by being my mother.

My youth gives me confidence that I will have many more love-filled relationships, but I doubt any will surpass the bond that my mother and I shared.

I've immortalized my mom with the reverence due angels such as the ones left behind in her bedroom. But I'm entitled. After all, she is my mother.

I pressed the video remote to let the *Larry King Show* play out. "Patti, you were a child when Sharon was killed. Do you still think about her?" King asked.

"All the time; and it hurts just as bad today as it did then."

"We've run out of time," King said, "but we obviously haven't heard the last of this case."

Nine years from the initial broadcast of that show, King's assessment of Van Houten's case blossomed and snared like a well-fed bougainvillea. Following her June 2000 parole denial, Van Houten filed a state appeal that accused the parole board of basing their decision solely on her commitment offense.

In order to deny parole under California statute, the board must present clear and convincing evidence that a prisoner would pose an unreasonable risk of danger to public safety. If the Superior Court found the board negligent of upholding the law, Van Houten would be entitled to a new hearing and possibly a release date.

The Superior Court set a hearing for May 23, 2002.

I followed the news headlines with interest because I wasn't sure if anyone would petition to keep Van Houten in prison. Aside from my brother and sister, my immediate family on the Tate side has all passed. There is my Aunt Debra. Though she is my mother's sister, she is a far-removed aunt. Never close with Mama, often at battle with her, I knew Debra less than I knew Aunt Sharon. And because we were such strangers to each other, I had no way of knowing if she cared enough to fight to keep Aunt Sharon's killers in prison. What little I knew about her firsthand was that she'd taken my grandfather's ashes and, to date, had not given him a proper burial. Under those circumstances I wasn't sure if she was the one to carry on my grandmother and mother's work.

Like a majority of the public, I lacked the facts of Van Houten's case; therefore, she easily swayed me with her humbleness. I was not alone. In 1969 Van Houten just as easily swayed detective Mike McGann when he interrogated her at the Inyo County jail. Van Houten toyed with McGann. Coy and shrewd, mixed with nervous laughter, Van Houten gave up little information on her or her crime partners. Instead, she steered McGann's attention to her alibi for the Gary Hinman, Cielo, and Waverly Drive murders. "For about three weeks I was laid up because of a knee injury. And I didn't go to the doctor, so for about three weeks, I was just sort of out of it."

"When was that?" McGann asked.

"End of summer. About the end of July, beginning of August. So for about four weeks I was out. I'd get up to go to the restroom and black out. I had to stay in bed because I couldn't walk, so I didn't know too much about what was going on."

"Leslie, tell me what you heard about the Tate murders up there."

"Not much."

"Here's the thing. Five people were killed up there. I know three for sure that went up. I think I know the forth, but I don't know the fifth—are your parents still alive?"

"Um, I couldn't tell you," Van Houten said.

"You don't have any feelings for them at all?"

"Oh, I love them but—"

"Would you want to see these people, for no reason at all, to go up there and kill your parents or brothers or sisters? Do you think that's right?"

"No."

"I don't either," McGann said, "but that's exactly what happened. And you know exactly who was there and what went down. I need to get at what you know, what you heard. That's all I'm asking."

"Well, I heard that Sharon Tate was murdered by four other people, and I heard PIG was written on the front door, just like they did at Hinman's—just like ole Sadie Mae did before. Um, I heard that the Folger girl was stabbed like eighty-seven times."

McGann interrupted. "Kind of vicious, huh?"

"I'll say. Yeah. Overdoing it a little. Anyway, that's about all I know."

Interpretations vary. Vincent Bugliosi included part of the Van Houten interrogation in his book *Helter Skelter*. It included asides such as laughs, sarcasm, shock, and anger, and he riddled her quotes with exclamation points. I interpreted this vast fluctuation as maniacal emotion. When I heard the tape, I formed a quite different opinion of Van Houten. She remained calmly equivocal throughout the interrogation. She did often giggle, but it was the nervous laugh of one about to be charged with murder.

Clearly Van Houten was a sane individual who wasn't worried about the end of the world because her last question to McGann was, "So, what's this grand jury immunity deal?"

The show *Both Sides* aired an hour before Court TV began the live coverage of Van Houten's next parole hearing. I watched, not only to familiarize myself with the details of Van Houten's participation, but also to learn the details of the LaBianca murders. Additionally, I wanted to ensure that someone was indeed representing the victims. A picture within picture popped up on the screen that showed the parole hearing room. My confidence was raised by the appearance of the LaBianca family members, coupled with Van Houten's shackled entrance. The restraints seemed a sure signal from the board that, until proven otherwise, she remained a threat to society.

Midsentence of one of the Court TV reporters, Board Commissioner Sharon Lawin silenced them by starting the hearing with a statement of facts concerning the murders of the LaBiancas, then asked, "Miss Van Houten, are you responsible for the deaths of Rosemary and Leno LaBianca?"

"Yes," Van Houten eagerly replied.

"Now, your case covers a longer period of time than August 10, 1969, in terms of your involvement and what led up to this particular action. The night before, multiple murders were committed by your crime partners and others. Now, you didn't participate in that, but as I read the records, you wanted to or felt left out. Would that be an accurate reflection?"

Van Houten nodded. "I found out the next day what had happened [at Cielo] and leading up to that point, Manson spoke a lot about sacrificing ourselves to the beginning of the Helter Skelter war. My loyalty and need to please him made me want to go the second night. Also, I kept very close to Patricia Krenwinkel. So when I realized that she had gone [the night before] it was important that I also go."

"Did you know beforehand that you were going to commit murder?"

"Yes. I knew at the ranch before we left."

"When you arrived at the LaBiancas'. Who went in first?"

"Manson, maybe Tex. I'm not sure."

"Did you stay in the car?"

"Yes."

"When Manson came out, what did he say?"

"He looked in the car window and told Pat Krenwinkel and I to get out. He said to make sure that everyone did something and he asked that the people not be frightened."

"When you went inside, where were Mr. and Mrs. LaBianca?"

"They were on the sofa, tied up."

"And then what happened?" Lawin asked.

"Tex told Pat and I to take Mrs. LaBianca into the bedroom, but first, Pat and I went to the kitchen and she got some knives, but I don't remember if she handed me one or not because I used both hands to hold Mrs. LaBianca down in the bedroom."

"Where was Tex Watson at that point?"

"He was in the living room killing Mr. LaBianca. I heard sounds of him dying, the gurgling sounds of him dying, and Mrs. LaBianca heard them, too, and started to struggle while calling out for her husband. Pat tried to stab her with a knife, but it bent. I couldn't hold Mrs. LaBianca down, so I called for Tex and told him that we couldn't kill her. When Tex came in, Pat left the bedroom, and I stood at the threshold and stared out into the hallway. I stayed there until Tex turned me around and said, 'Do something.' When I looked in, Mrs. LaBianca was lying on the floor dead. I stabbed her in the lower torso approximately fourteen to sixteen times. After that, I told Pat that I had touched the lamp, so I began wiping off fingerprints in the bedroom. And I wiped off fingerprints in there for as long as I could."

"At some point, you went looking for a change of clothes?" Lawin asked.

"Yes. I had no blood on me, but Tex had showered in the bathroom and needed the pants that I had worn. So I went through her closet and found a pair of shorts to change into."

"Anything else to tell us about the commitment offense?"

"I'm deeply ashamed of it. Living with the acts of that night is difficult. I take them very seriously, not just the murders in the house, but what was in me that made me so available to someone like Manson. One of the hardest things I have to deal with in contributing to murder is that there is no restitution. There's no making it right. My heart aches with words, but there don't seem to be any that can really convey living with the amount of pain caused."

Lawin studied her briefly. "You said a moment ago that you contributed to murder. Now, did you *murder* Mrs. LaBianca or *contribute* to her death?"

"I feel I contributed to her death," Van Houten said, and then added, "It's difficult to answer that because the autopsy reports have shown that it was Tex that wielded the fatal wounds, but I contributed. I attempted to hold her down for Pat, and I called Tex because we couldn't kill her. Morally, I feel as though I did."

"And are you certain that she was dead when you stabbed her?"

"I felt she was. I didn't think in terms of absolutes at the time."

At the first commercial break, I pulled out the death photos that I had evaded for nearly ten years. I still couldn't bear to look at Sharon's, so I shuffled through them until I found the pictures taken at the LaBianca house. I tried to review them clinically in order to compare Van Houten's G-rated rendition to actuality, yet my pulse quickened. Mr. LaBianca lay wedged between a chair and sofa. A bloodied pillowcase covered the harrowed look that likely remained on his face; stab wounds covered his body. The killers had ripped open his blood-soaked pajama top so they could gouge WAR into the flesh. They left a fork embedded in his abdomen, a knife traversing his throat.

Mrs. LaBianca lay on the bedroom floor also with a bloodied pillowcase cinched around her head by a lamp cord—presumably the reason Van Houten "touched" the lamp. Her buttocks and back were exposed because the killers had yanked her gown up around her shoulders. Deep stab wounds riddled her backside. Van Houten admits this is where she reluctantly stabbed Rosemary, yet the photographs of these wounds clearly indicate they were not inflicted by an unwilling hand.

Based on Van Houten's interview with King, and her testimony at the parole hearing, *honest, nice, sad,* and *remorseful* are all adjectives I could use to describe this woman. If Van Houten is a culmination of those adjectives now, how important is it to remember that she stayed in the LaBianca house long after her victims were dead? That even as the blood still seeped from Leno and Rosemary, Van Houten and her crime partners showered, ransacked closets, raided the refrigerator, and—as indicated by the blood-matted fur—played with the LaBiancas' dogs? I think the apex of that question lay in

Dr. McDaniel, the psychiatrist who most recently examined Van Houten: "Miss Van Houten possesses a degree of personal charm that is very convincing. The obvious question is whether this represents true emotional change and a restructuring of her personality, or of someone who is so smooth in their manipulations that they are barely perceptible."

The doctor's supposition was reason enough for the board to deny Van Houten parole for another two years.

In the aftermath of the six-hour hearing, I kicked around the parallel course of Van Houten and Sharon's early lives. Both were homecoming princesses and popular with their friends. Both made high grades. Both experimented with drugs. How they ended up on such diverse paths is a mystery that psychologists should be given carte blanche to solve.

Nana and Mama tangled with forgiveness of Sharon's killers until they were hog-tied. I, too, have irresolvable issues on the subject. Essentially, parole equals forgiveness. Forgiveness comes once a prisoner fully acknowledges what they did. Furthermore, forgiveness comes once they've made restitution to their victim. Since it is impossible for a killer to make restitution to their victim, I believe that murder is unforgivable. Consequently, no matter how rehabilitated they are, they should remain imprisoned for the rest of their lives.

In the wake of every parole hearing Nana attended, press members asked her how much longer she would continue her opposition. She would say, "Until the day I die, and then some."

Of course, things have changed at the parole hearings since Nana's time. Steve Kay retired; members of the LaBianca and Sebring families have stepped forward to voice their opposition to the murderers' release; and Proposition 9 now allows the board to give prisoners up to a fifteen-year denial. To date, the parole board has not released any of the Manson Family killers, although, in a plea bargain of sorts, Steven Grogan received parole after revealing to the state where they could find the long-lost remains of Shorty O'Shea. Grogan has been living carefree in Northern California, under an assumed name ever since.

But it's hard to forget that while he never entered the house that night, he waited in the getaway car. Atkins and Kasabian were charged with two counts of conspiracy to commit murder, which, in the eyes of the law holds the same consequence as murder. It remains unclear as to why Grogan was overlooked. But there is no statute of limitations for murder. Every day that the District Attorney's office remains blind to their oversight, Grogan should count his undeserved blessings.

The Parole Board did give us a scare in 2010 when they granted Manson killer Bruce Davis a date for release. Thankfully, saner minds prevailed, and Governor Schwarzenegger invoked his authority to reverse the board's decision.

The year Nana passed away, the California Department of Corrections removed Charles "Tex" Watson from his homey environment at the Men's Colony. These days he resides at Mule Creek State Prison, divorced from his wife and disrobed from his ministerial duties. Apparently, he's doing an outstanding job as one of the prison janitors.

Time has come full circle since Nana began her fight to keep Sharon's killers in prison. At the time, due to overcrowded prisons, the parole board was granting a high number of releases. Nana and the multitude of other victims' advocates became the guardian angels of the state as they worked tirelessly to keep the violent offenders behind bars. They did their job right, but California didn't, as the state continually neglected to build more prisons to house its offenders.

Now, nearly thirty years later, due to the same overcrowding issues, the U.S. Supreme Court has mandated that California release 40,000 prisoners over the next two years. It wasn't the answer then and it's not the answer now. But on November 16, 2011, Watson saw the mandate as a crack in the prison door and his best chance of getting a release date. During his five-hour parole hearing, Watson read a statement regarding his suitability and then refused to answer any questions from the board members. Because the board measures an inmate's suitability based in part on that inmate's response to their questions, it was a bad call by Watson and the crack in the door snapped shut on him with a denial that will keep him in prison for a minimum of five years more and quite possibly forever.

After years of flying under the radar, Patricia Krenwinkel's 1993 parole hearing was perhaps the most damaging hearing she's had to date, and I like to think that my mother played a large role in her eventual meltdown that day. She was the first victim to ever confront Krenwinkel and had brought forty-two thousand petitions to support her effort. By this time, Mom was no longer the frightened, timid person she'd been at her first hearing. She'd attended three other hearings for Sharon's killers, and she was a force to be reckoned with.

There were other key players at the hearing: Steve Kay, and Tom Giaquinto, the commissioner of the Board of Prison Terms. As an ex–homicide detective, Giaquinto was trained in the most effective ways to interview a witness, and for the next hour, he grilled Krenwinkel about the murders in what I believe was an attempt to surface her repressed hostility.

Krenwinkel's answers to how seven human beings were eliminated from this world at her hand were bland at best—or as she described herself in 1969, "very dead inside, empty, and just completely kind of hollow." I think it was this attitude that incited my mother's anger by the time she made her impact speech and said, "I feel like she has no memory of what happened that night, and what she did. She didn't know them . . . I did, I knew Gibbie. Does she remember what she looked like? . . . Does she remember when she led my sister down the hallway to her death, what she looked like, with her belly out to here? She doesn't hardly remember what she did. And I want to put some faces and feelings here because I feel like she's just so blank."

Through the course of the hearing, Giaquinto threw question after question challenging Krenwinkel's participation, but in the end, it was board member Guaderrama's question that flung her over the edge: "How do you feel now about what you've done?"

As she responded, her composure unraveled more with every word until she was just a sobbing heap of outrage and anger. By the time she got out her last sentence, she had turned her animosity toward Mom:

It was as if she was incensed that Mom wasn't more sympathetic toward her and how difficult her life had become since the murders. "It's grotesque. It's absolutely horrible. It's very difficult for me to live with the fact that I could do anything so horrible. . . . So I try every day, the best I can to deal with it. Every day of my life, I try to define to myself that I have a little bit of self-worth because it is terribly difficult to deal with this. . . . No matter what I do, I cannot change one minute of my life. And, as I said before, I don't expect the board to say that I can go home, I am paying for this as best as I can. *There's nothing more I can do outside of being dead to pay for this and I feel that's what you wish, but I cannot take my own life.*"

Mission accomplished. The hearing ended soon after with a three-year denial.

In an interview with Court TV following the hearing, Krenwinkel's resentment of Mom's presence was apparent when she said, "After they removed the death penalty, I have been granted the right to return to a parole board for *them* to consider within *their* hearts and *their* judgment whether or not I fit a criteria. . . . So it's up to *them*, no one else."

Perhaps, but as with the parole board, Court TV gave Mom the last word on the subject, "To sit back and judge the taking of a life by the amount of years spent by that prisoner is not good enough for me. The prison system is set up for many purposes. Retribution is one of them, but the biggest is safety, isn't it? The biggest issue is to not make more victims."

HIGH COURT SPURNS LESLIE VAN HOUTEN'S BID FOR RELEASE, read the headlines for the Supreme Court's final ruling on her case. Presiding Supreme Court Justice Manuel Ramirez wrote in the record words like "cruel and callous" before stating, "The courts must uphold the board's exercise of its discretion to find an inmate unsuitable for parole as long as there is 'some evidence' to support it. And that evidence can come solely from a review of the circumstances of the crime."

And that's just what happened on July 6, 2010, when Van Houten had yet another parole hearing. As if flaunting the Supreme Court's

2002 ruling, Chairman Robert Doyle noted in their decision for denial, "The crimes involved were so atrocious and heinous that they
must be considered in the decision."

Of course, they cited other reasons for Van Houten's unsuitabil-
ity. Most noteworthy, they stated that even after forty-two years, Van
Houten has "failed to gain complete insight into her crime and its
motivation. . . . She does not look at herself to see what made her capable of this activity."

Whereas in the past, Van Houten had received a one- or two-year
denial, this time, they gave her a three-year denial.

*Twitter is the frontier, the wild west of cyberspace where outlaws, gu-
rus, the lonely, the incarcerated can communicate with YOU.* The fact
that Charles Manson was able to communicate that message to a
forum of millions from the confines of his Corcoran prison cell is extremely disturbing. Cell phones in prisons have become an epidemic.
At Corcoran, in just one day, officials were able to track more than five
hundred cell phones and four thousand attempts by prisoners to text,
make calls, or to get an Internet connection.

In 1969 Manson had maybe a hundred followers. Today, courtesy
of a cell phone and his Twitter account, he has more than five thousand followers, including Lynette "Squeaky" Fromme and convicted
killer Bobby Beausoleil.

At the moment, all the lobbying in the world can't stop this because
California doesn't have the funding to get it under control, and even
if they managed the funding, there's no law to support the cause. In
California, it is not illegal to smuggle a cell phone into a state prison.

The CDC did make an effort to thwart cell phone use when they
launched "Operation Disconnect," in which they conduct surprise in-
spections looking for phones. On January 6, 2011, they confiscated a
cell phone from Manson. By January 7 he was back online and tweet-
ing. Considering the multitude of infractions he has on file, the thirty
days they added to his sentence is hardly a deterrent.

For Manson, a cell phone infraction may be a moot point if he wins
a new trial. From the moment of his conviction, Manson has claimed

that he didn't get a fair trial because he was not allowed to represent himself nor put on a defense. With the help of Saddam Hussein's former attorney, "The Devil's Advocate" Giovanni Di Stefano, Manson has filed a petition with the Inter-American Commission on Human Rights. In an interview, Di Stefano commented, "I have no interest in the facts of this case. The law is the law. Denying his rights under the Sixth Amendment to represent himself is grounds for a new trial."

And so the saga continues.

Papa was right when he said Linda Kasabian was as guilty as those imprisoned. For me, her name is synonymous with *what if?* What if she'd run to get help? What if she'd gone to the police? The oft forgotten Linda Kasabian has managed to remain out of prison, but not out of trouble. She has been arrested on at least two occasions; once in her apartment filled with guns and narcotics, once for disturbing the peace. In 1989 she reemerged to the spotlight for a *Current Affair* interview. On the show, Kasabian recalled her crime partners with a wistful gleam and a smile. Of Krenwinkel she said, "Quiet, earthy." Of Atkins, she laughed heartily, "She really knew how to talk. She knew all the right words." And of Watson, "Tall, dark, and handsome with deep, deep blue eyes."

For someone who was supposedly so horrified and remorseful, these are very peculiar adjectives for her to use considering the last time she was with any of them in free society she watched them relentlessly stabbing their victims.

I first laid eyes on Susan Atkins when I attended her 2005 parole hearing. That day I had bronchitis and a 102 fever, but I'd gone anyway, intent to make an impact statement against the woman who had brought so much pain to my family. For reasons that are still unclear, I was denied that right. Debra Tate spoke instead. At that point, given my condition and the fact that I had been silenced, I should have gone home; yet I still hoped that I would catch Atkins's eye, if even for a second, to let her know that her horrible actions had flowed through and impacted another generation. So I stayed and endured.

In California, by definition of the penal code, the reason a defendant is incarcerated is for punishment. There's nothing about rehabilitation, and yet I had to listen to four hours of discussion about how her rehabilitation entitled her to a parole date. That was followed up with an endless list of organizations and self-help groups she'd participated in. Even in my weakened condition, it was tough to suppress a laugh by the time the born-again Christian got to the Shalom Sisterhood.

In all, there are only two segments that are noteworthy. The first came when Atkins's husband/attorney, James Whitehouse, made a note at the end of his closing statement: "When they (the victims' next of kin) make their statements, they say that we've never tried to apologize. And I don't think they realize that the Board of Prison Terms and the CDC won't allow us to do so."

Presiding Commissioner Perez interrupted him, "That's inaccurate, sir. . . . You may, but there is a process to follow." And then, a moment later she told Atkins, "This is your opportunity to provide a statement to the panel relative to your suitability for parole. If in your statement you wish to indicate that you are sorry for what you've done to the family as well as the victims, you may do so; however, you have to address them through the panel."

Nana must have been looking down on this scene, waiting on pins and needles for her reply. It seemed that after four decades Atkins was finally going to utter the words, I'm sorry. . . . Only, she didn't. Even after her husband made it a point of contention, Atkins couldn't seem to bring herself to say those two simple words. Instead, she chose to use her opportunity to read letters from those supporting her release.

The other significant moment came directly after that when the panel turned the floor over to Jay Sebring's sister. In under a minute, she countered the preceding four hours with a simple summary: "She's got all these things she's done, all these committees, all these organizations, all these degrees, but the bottom line is, she's just as much a murderer today as she was thirty-six years ago because my brother is just as dead as he was thirty-six years ago."

And with that, I knew Nana had said a quiet *amen*.

In the spring of 2008 Atkins was diagnosed with a cancerous brain tumor. She and her family appealed for compassionate release from prison so that she could die in the comfort of her family's home.

At present there are more than eleven thousand elderly prisoners in California, and the Manson women are among them. For anyone with a heart, it's tough not to feel sympathy when prisoners' rights advocates who support compassionate release show us pictures of old, fragile, often sick women and say, "It's a terrible injustice . . . When you're in prison, all you want is to be able to die with dignity . . . I saw a woman get down to her knees and beg for morphine . . . I saw one woman with throat cancer, who kept getting denied parole, fall into her own blood and die."

While my heart aches for any suffering human being, I forced myself to look away from photos of the pathetic and dying Susan Atkins and remember the young and vibrant Atkins who with callous disregard unjustly murdered eight people and watched them beg for their lives, fall down into their own blood, and die without an ounce of dignity. The parole board must have agreed because at her last hearing in 2009, with full knowledge that she had only months to live, they gave her a three-year denial.

On September 24, 2009, Atkins's husband, James Whitehouse released this statement: "Susan passed away peacefully surrounded by friends and loved ones . . . Her last whispered word was 'Amen.' "

Well, God bless her, because that's more than I can say for any one of her victims.

Not a day goes by that I don't think about Aunt Sharon, Nana, Papa, and Mama. With them, I often contemplate the word *fate*. Despite Mama's wish to have our name disentangled from Charles Manson's, I'm left with no choice but to integrate the names, because on August 9, 1969, in an hour's time, Sharon's killers played with fate and changed destiny's blueprint until theirs merged with ours. Unfortunately, that night, those killers and their actions became part of who I am.

Just as Sharon was the night of her murder, I am almost eight months pregnant. As I wait for my baby to arrive, I spend every day planning for our future together. As her arrival draws nearer, I shop for our needs and ready the house. I spend most days in the nursery; first painting, then furnishing, and now, lovingly placing mobiles, photos, bedding, and organizing clothing in the drawers. Every morning, I awake excited about our future. And every night, as I drift off to sleep, I can't help but think about the night Sharon was murdered, because as she safely got into bed, she must have been having the same thoughts and feelings that I'm experiencing. In these darkened moments, alone in bed, with the baby that I feel so much love for; the baby that I've named Patricia, who is already full of life and constantly shifting in my belly as she readies for her birthday, I finally, truly comprehend how Sharon must have felt as she pleaded with her killers, "Just let me have my baby and then you can kill me any way you want." As I'm sure it was for Sharon, my baby is my world and I would gladly give my life to save her life. Each night, as my heart soars with joy, it also breaks for Sharon's loss. My last conscious thought before dreams overtake my mind is a prayer that I will safely awaken to have at least another day with my child and hopefully a lifetime together.

Manson, Watson, Atkins, Krenwinkel, Van Houten, and Kasabian are thieves as much as they are killers. They stole Sharon's future, and in turn, stole the future of every generation to follow, leaving us with the infinite question of, what if?

I nostalgically roam the quieted rooms of my house where three generations have passed through the portals of time and a fourth generation is about to begin. Everywhere I go, I'm pleasantly haunted by days gone by. It seems these walls do talk with remembrances of a lot of joy, a lot of love, and a lot of sorrow.

Sorrow may seem a pitiful emotion, yet I am blessed by it. For it has left me a legacy of love, determination, and courage handed down by my mother and my grandparents before her. They taught me that we learn our greatest lessons through hardship. And through that hardship, they taught me not about fear and retribution, but about giving.

In the last year of Mama's life, I went to bed every night searching for one special gift that I could give her in which only she benefited.

I'm still searching. Because I have yet to find a gift in which I won't reap in the blessing of giving unconditionally.

I am enamored by Father O'Reilly's eulogy for Sharon: "We create in every act of good we do; we destroy in every act of evil we perform." For in it is the key to a civilized society much like Catherine Ryan Hyde's novel *Pay It Forward*, about a boy who believes in the goodness of human nature and sets out to change the world with one act of kindness. An act, which instead of being paid back to him, is paid forward to three others, and so forth.

Like a stone pitched into a dead calm, every act committed toward another—good or evil—will reach a hundred more by the day's close, and in some cases, like Aunt Sharon's, millions are affected. Mom and Nana spent their lives throwing positive stones. The outcome of their ripples will remain a mystery, but I like to believe that their splash is still circling outward and paid forward by those who don't even know where the epicenter formed.

President Bush honored Nana, but Mama left this world clueless (and in denial) to the beneficial impact from her stones. Many of the issues she challenged didn't come to fruition until after she passed, such as Manson's song being removed from the Guns N' Roses CD, or the abolishment of conjugal visits for life-sentenced prisoners. Moreover, through her guidance, we, her children, matured into responsible, caring adults. I hope now she understands the lasting impression she made during the short time she was here.

My grandparents and mother left me a well-cultivated garden of endless growing capabilities. I owe it to them—at the very least—not to cast any negative stones. As I tend to that garden and plant my own seeds, perhaps one day, as I dig through the soil, I will uncover and collect a handful of contributive stones and pay it forward.

In my path, I find the dimes that Mama promised to drop from heaven—inflation, she joked. At least once a week I find a dime that reminds me that she's still close at hand, watching over me; keeping me alert that at any moment her spirit could startle me with a quiet "boo."

I think about the past and I think about the future. Both seem so distant yet so close; according to some, they run simultaneously.

I believe that with the final court denial of Susan Atkins's plea for compassionate release from prison, it's a safe bet that the rest of Sharon's killers will never leave prison alive. The parole hearings will continue, but I believe that the need to attend those hearings to give a victim's impact statement has finally been laid to rest with my mother's generation. If that situation changes, you can be certain that I will be the first one in line to oppose their release because, I figure I'm Nana's and then some.

If you or someone you know is a victim and in need of support, please contact Crime Victims United at Crimevictimsunited.com, 530-885-9544, mail@crimevictimsunited.com.

All victims organizations are wonderfully productive and helpful. We listed Crime Victims United here because the president/chair, Harriet Solarno, and vice chair, Marcella Leach, were both good friends of Doris Tate's and mentors to Patti Tate. Patti remains an executive honorary member of CVU.

And of course this book would not be complete without contact information for Parents of Murdered Children. 1-888-818-POMC, www.pomc.com.